The Evolution of Earth's Climate

Scrivener Publishing
100 Cummings Center, Suite 541J
Beverly, MA 01915-6106

Publishers at Scrivener
Martin Scrivener (martin@scrivenerpublishing.com)
Phillip Carmical (pcarmical@scrivenerpublishing.com)

The Evolution of Earth's Climate

J. O. Robertson, G. V. Chilingar,
O. G. Sorokhtin, N. O. Sorokhtin,
W. Long

Scrivener
Publishing

WILEY

This edition first published 2018 by John Wiley & Sons, Inc., 111 River Street, Hoboken, NJ 07030, USA and Scrivener Publishing LLC, 100 Cummings Center, Suite 541J, Beverly, MA 01915, USA

For more information about Scrivener publications please visit www.scrivenerpublishing.com.

Wiley Global Headquarters
111 River Street, Hoboken, NJ 07030, USA

For details of our global editorial offices, customer services, and more information about Wiley products visit us at www.wiley.com.

Library of Congress Cataloging-in-Publication Data

ISBN 978-1-119-40706-5

Cover image: Cloud and Earth View, Jorge Villalba | Dreamstime.com
Cover design by Kris Hackerott

Set in size of 11pt and Minion Pro by Exeter Premedia Services Private Ltd., Chennai, India

10 9 8 7 6 5 4 3 2 1

Dedication

This book is dedicated to those politicians interested in the truth, basing their opinions on scientific facts, rather than emotions, personal profit, or conformity. An excellent example of this is the United States President. Donald J. Trump

Dedication

This book is dedicated to President Fu Chengyu for his support and outstanding contributions to the global Petroleum Industry and as an outstanding world leader in the development of energy.

Dedication

This book is also dedicated to the memory of Dr. O. G. Sorokhtin and Dr. Misha Gorunkel, their invaluable help, and to Dr. John Mork for his tireless support of the University of Southern California and his help in publishing this book.

Contents

Introduction

It is time for us to "let science be science!"

Unfortunately, today one often sees some political and scientific leaders making climatic decisions based upon political conformity, the desire for personal profit, and/or emotions expressing what they want it to be and not what it is. These decisions often misstate scientific facts and label their opinions as scientific fact. This is particularly true in climatology, in discussions of global warming resulting from man's activities. Over the past few years, one has seen ill-informed individuals making catastrophic statements about how global climate change will damage the Earth. These are often individuals who: (1) have a partial understanding of only a limited aspect of the natural sciences, (2) ignore long-term historical scientific data, tending to only look at the very limited current data of the past few years, (3) support their decisions because they can make money selling snake-oil to others, and/or (4) often appear to be more interested in making it a political issue and refusing to seriously discuss it. For all the above reasons, and possibly a few others, one can see politicians and a few scientists making foolish predictions about how climate will drastically change in the near future—predictions which do not prove to be true—and then calling their proclamations science. The authors refer to this as ***pseudoscience***, in which people who predict catastrophe refuse open discussion of their ideas, unless we are willing to accept their ideas. This same group of individuals encourages politicization of science while discouraging any discussion that might not be acceptable to their perceptions.

A geologist Robert S. Young (2017) wrote in the *New York Times* about science marches supporting *anthropogenic* global warming,which was later echoed by many on social media, "a march reinforces the narrative that scientists are a special interest group, and further alienates scientists and scientific findings from a significant portion of the population who are already skeptical. Will a march for science help science?" It seems unlikely.

The fact that over the past 50 years many projections and predictions have proven to be false, has never changed their unsupportable forecasts.

Early scientific leaders, Galileo, Newton, etc., developed what is today called the *scientific method* in that one: (1) first gathers the data, (2) then develops a hypothesis to explain the gathered data, (3) tests that hypotheses with data and then (4) if the hypothesis doesn't totally answer the question, *adjusts the hypothesis* and continues testing until all aspects of the hypothesis are covered. There is open discussion during all aspects of this method. Unfortunately, the pseudoscientists develop their hypothesis and then refuse to discuss their concepts, politicizing and demonizing anyone with a question that challenges their hypothesis, even though there is data that conflicts with their hypothesis.

This book is dedicated to use of the scientific method in explaining the development of the Earth's climate and addressing the questions, "(1) *Is there global warming? and (2) is it generated by human beings?*" To do this requires an understanding of data from the fields of geology, thermodynamics, physics, cosmology, biology and other areas of natural science. As additional scientific data is revealed, the hypotheses presented in this book will change accordingly.

The followers of pseudoscience have proposed many catastrophic changes in the Earth's climate over the past few years, which have been proven to be false. Friends of Science (2017) have succinctly grouped many of these *misconceptions* into the following myths:

1. Global temperatures are rising at a rapid, unprecedented rate.
2. The *hockey stick* graph shows that the Earth has experienced a steady, very gradual temperature decrease for 1,000 years, then recently began a sudden increase in temperature due to increased *anthropogenic* carbon dioxide in the atmosphere.
3. Human-produced carbon dioxide has increased over the last 100 years, adding to the *greenhouse effect*, thus causing most of the Earth's warming of the last 100 years.
4. Carbon dioxide is a pollutant and a dangerous poisonous gas in the Earth's atmosphere.
5. Computer models verify that increases in carbon dioxide in the atmosphere cause significant global warming.
6. The United Nations Intergovernmental Panel on Climate Change (IPCC) has demonstrated that man-made carbon dioxide causes global warming.

7. Global warming will cause more storms, fires, rising sea levels and other catastrophic climatic extremes.
8. Receding glaciers and the calving of ice shelves are proof of anthropogenic global warming.
9. The Earth's poles are warming, polar ice caps are breaking up and melting, and polar bears are dying out due to global warming.

Every one of the above statements is false, proven by scientific facts. Yet the proponents of this pseudoscience continue to make additional outrageous climate forecasts, supported not by science, but rather their feelings.

The authors have gathered the following scientific data to present a factual understanding of the ***Evolution of the Earth's Climate.*** The authors for this book present the collective wisdom of many authors with experience and expertise in the fields of geology, engineering, mathematics, cosmology, biology, thermodynamics. As more scientific data is gathered, the hypothesis within this book will change; that is science!

Acknowledgments

This book is the result of the work of many contributors whose research has enabled the writers to build upon their work. The late O. G. Sorokhtin's notes were made available by his son, N. O. Sorokhtin, enriching the concepts presented in this book. The authors have attempted to give credit for the many ideas and research of many previous investigators. Without their ideas, this book would have been impossible. Science is an accumulation of ideas that slowly grows over time. Our understanding of the present is only accomplished by understanding the past. Hopefully, as our base of knowledge continues to grow, the ideas expressed in this book will be again modified to match the scientific data of tomorrow.

The writers owe thanks to the editor, Dr. Phil Carmical, and his staff who made everything presented a little - and sometimes at lot - better. Special thanks go to Dr. Misa Gorfunkel for his help

Academician John O. Robertson owes a special acknowledgment to his wife, Karen Marie Robertson, for her support over the past 50 years and to his son, Jerry J. Robertson, for his important contributions. Thanks, is also extended to Joel Hinrichs and Doug Praska for their thoughtful comments in reviewing portions of the book.

Professor, Academician George V. Chilingar wishes to acknowledge the moral support and encouragement given by his wife Yelba Maria Chilingar and his children Eleanor Elizabeth, Modesto George and Mark Steven.

Mr. Wennan Long is very thankful for the support of his parents, Mr. Guowen and Mrs. Chen Long; and his wife, Mrs. Lewei Long.

Part I
CLIMATIC PARADOX

1

Climatic Paradox

Historic Temperatures of Early Earth

Today, the Earth is the one planet out of all the planets in the solar system that has an atmosphere providing a favorable climate for the evolution of life (both flora and fauna) on its surface. To better understand the historic development of the Earth and its atmosphere, geologists have divided the age of the Earth into eons, eras and periods. Figure 1.1 displays the geologic system of time division used in this book, showing the development of life and how it relates to the changing environmental geologic development of the Earth. When discussing the climate of the Earth, a critical component is that of temperature. A temperature chart covering the entire history of Earth is presented in Figure 1.2. The arrow on the right side of this figure points to today's average surface temperature of the Earth. The Earth has experienced both cooler and warmer periods of temperature with cyclic occurrence. In general, Figure 1.2 reveals that the Earth has been much warmer in the past than it is today, and that over the past 65 million years (MY) has progressively cooled.

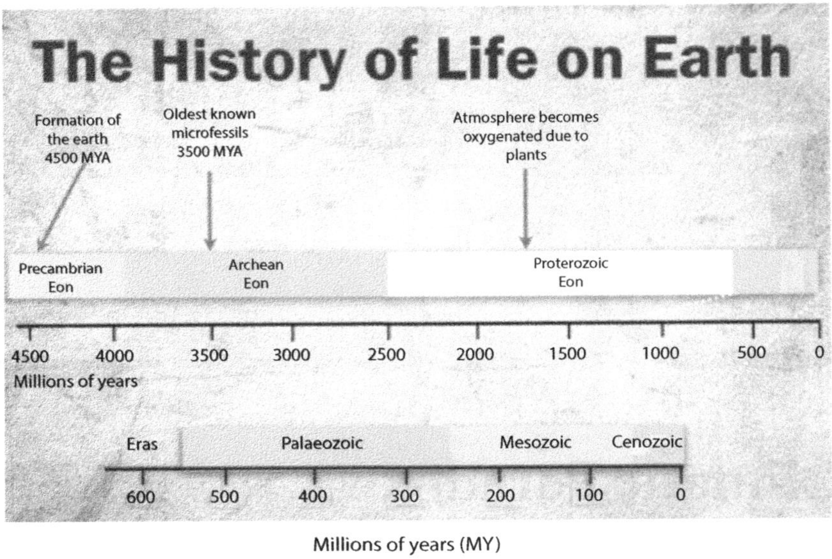

Figure 1.1 Development of life on Earth for the past 4.5 BY. (After pbhslifescience, https://pbhslifescience.wordpress.com/2016/11/04/evolution-slides-1.)

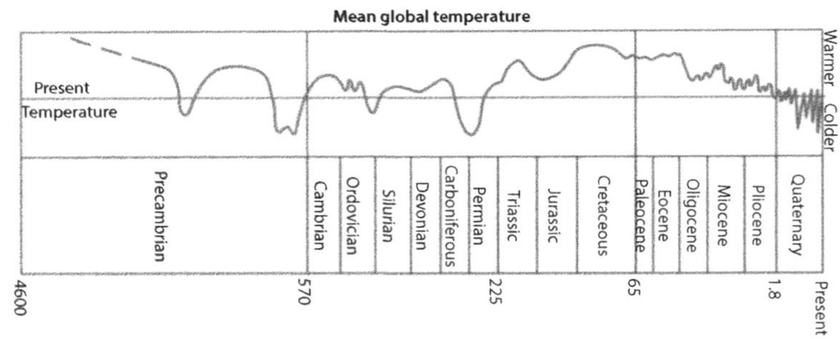

Figure 1.2 Schematic 4.5 BY Earth chart. (Original data from Frakes (1979); in: Watts, *https:// Wattsvpwiththat.com/2014/09/08/Monday.*)

Sorokhtin *et al.* (2011) defined life on the Earth as a concourse of several circumstances. These circumstances include: (1) a quiet star, e.g., Sun; (2) the optimum distance of the Earth to the Sun throughout the Earth's orbit about the Sun; (3) the presence of a massive satellite, e.g., the moon; (4) chemical composition of the primordial Earth and its atmosphere; and (5) several other causes reviewed in detail in the monograph by Sorokhtin and Ushakov (2002). Climatic conditions for life habitation on the Earth was a result of changing atmospheric composition (see Figure 1.3) and pressure and to a feedback between the evolution of the Earth biota and atmosphere.

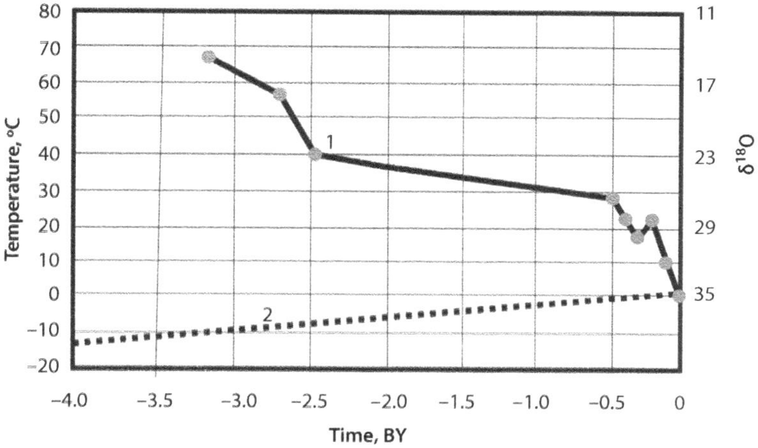

Figure 1.3 A climatic paradox: The sun luminosity increases over time, whereas Earth's temperature gets cooler. *Curve 1* – average isotopic water temperature that is sea-flint forming in, also describes the average temperature for the global, near-bottom of sea floor ocean water. The dots are $\delta^{18}O$ values for the sea-flint (after Schopf, 1982). *Curve 2* – temperature of an absolutely black body at the distance of Earth-from-sun, which describes the sun's luminosity.

Concepts by Some of Global Warming

Today by some, a major environmental concern is the release of human generated carbon to the atmosphere (anthropogenic). It is thought by these individuals, without either historic or scientific evidence, that humans are putting too much carbon in the form of carbon dioxide (CO_2) or methane (CH_4) into the atmosphere and that this anthropogenic carbon is somehow responsible for what they feel is *global warming.*

The Earth's climate (temperature) is primarily determined by: (1) the Sun's radiation intensity, (2) the composition and density of the Earth's atmosphere and (3) the Earth's atmospheric capability to absorb and retain the Sun's energy. It is often, incorrectly assumed, that the heating of the Earth's atmosphere and surface is, to a substantial extent, only due to the capacity of various atmospheric gases that absorb the IR (heat) radiation from the Sun, ignoring the atmospheric pressure.

The concept that the Earth's atmosphere is heated by the greenhouse gases was first proposed by the Swedish scientist Svante Arrhenius at the end of the nineteenth century. This idea is now postulated as obvious and accepted without verification [Maslin (2004) in: Enzler (2015); Budyko (1974 & 2002); Greenhouse effect (1989); and Petrosyants (1994)]. Svante proposed a relationship between atmospheric carbon dioxide concentrations in the atmosphere and global temperature, i.e., if only one increased the concentration of carbon dioxide, the atmosphere would absorb more energy.

Since then, this concept today is found within the conclusions of the Intergovernmental Panel on Climate Change (IPCC), Greenpeace, United Nations Environmental Program (UNEP), and World Meteorological Organization (WMO). This same viewpoint was totally supported in the resolutions of the International Ecological Congresses in Rio de Janeiro, Brazil (1992) and Kyoto, Japan (1997).

It is assumed that the main danger to the Earth's climate arises from the fundamental energy sources of the modern economy through the release of greenhouse gases, especially carbon dioxide, into the atmosphere. Under some forecasts of the proponents of the manmade (anthropogenic) *global warming* concept, a warming of the Earth's surface temperature could increase 2.5 to 5 °C by the year 2100. The rise in the ocean level caused by this warming would reach 0.6 to 1.0 m, resulting in problems for the densely populated areas of the Earth along the continental shores. Other consequences of *global warming* have also been forecast, e.g., the spread of deserts, permafrost melting, soil erosion, etc.

The conventual theory of *global warming* states that the heating of the atmosphere's troposphere occurs as a direct result of the anthropogenic addition of carbon dioxide (CO_2) and the methane (CH_4) that is added to the Earth's atmosphere by man. As a result, the petroleum industry has often been *singled out for blame,* for the release of additional carbon (methane and carbon dioxide) into the atmosphere as it produces, handles and burns hydrocarbons. Some of these environmentalists also believe that today's increase in atmospheric carbon dioxide can be primarily attributed to any human activities that are related to the release of CO_2 or CH_4 to the atmosphere.

Earth's Historic Temperature Charts

Today's literature is filled with a wide variety of options about global warming and climate change. Unfortunately, these opinions are not always grounded in the Earth's temperature history or scientific facts, but rather in emotional feelings. This range of opinions include those by: (1) The National Oceanic and Atmosphere Administration (NOAA) which has claimed that *there has been no recent slowdown in global warming*; (2) the Space and Science Research Corporation stating that *the Earth is presently in a sustained phase of global cooling*; and (3) some experts who state that *we have been in a transition phase without significant warming or cooling for the past 20 years.* These opinions are made without looking at the historic temperature charts of the Earth, leaving open the question of "***Are we cooling, transitioning, or warming?***"

Misuse of Temperature Charts

To correctly answer the question of the *Earth's cooling or warming controversy*, one must first look at temperature charts over the whole time of the Earth's geologic history to understand where we have been and where are we going. Understanding where we have been requires obtaining temperature charts (records) of temperatures before recorded history. Unfortunately, before recorded history, there was no recording of temperatures; however, there are several methods we can use to infer temperatures during these earlier periods of time. Temperature charts can be prepared by inferring temperatures obtained from ice cores, tree rings, etc. It should be noted that these values of temperatures obtained are approximations and not necessarily the exact values of the Earth's average surface temperatures (see Figure 1.2). Therefore, temperature charts, with values obtained by one method, may not always exactly match those temperatures obtained by a different method for the same period. Whenever possible, in preparing a temperature chart, it is preferable to use only temperatures that have been obtained by one technique. It has been said, *Oranges are easily comparable to other oranges, but not necessarily when you directly compare apples to oranges.* Also, many times it is not the inferred values of temperature that are important, but rather, the trends that a temperature chart data displays for a given period of time. This is particularly true when you have inferred temperatures from more than one method indicating slightly different numbers, but showing a similar trend.

Today, some portions of our planet experience extreme heat, whereas other areas experience the opposite. Temperature inferred at the poles should be expected to be cooler than the average surface temperatures inferred in other areas of the Earth and, therefore, not necessarily be expected to represent an average surface temperature for the Earth at that time. It has also been noted that climatic conditions of the two poles are not always similar. In examining a temperature chart, it is crucial to: (1) know what method was utilized to determine the temperatures and (2) whether in preparation of the chart, only one method was used to determine the temperatures.

Use of Paleoclimatology to Estimate Prehistoric Temperatures

Paleoclimatology is the study of historic climatic conditions, examining their causes and effects in the geologic past, using current data found in glacial deposits, fossils, sediments, etc. This chapter reviews the study of historic climatic conditions (temperature charts) of the Earth's temperatures as inferred by a particular method, e.g., analyzing ocean sediments. One should remember that before recorded history, neither satellites nor thermometers existed. But one may use the fossil record, for example, to infer/deduce temperatures at the time when the fossil of that flora or fauna was alive. For example, palm trees do not grow well in Arctic weather; therefore, the existence of palm tree fossils indicates a temperature at the time that fossil was laid down. Therefore, looking at fossils within the rock, we can infer an estimated temperature at that period of time the sediment was laid down, although the temperature may not be the exact value.

Estimates of the Earth's surface temperatures can be made by many different methods based upon the data from oxygen isotope ratios, ice cores, ocean sediments, dendroclimatology, etc. The various methods of temperature inference can result in slightly different values for the approximate temperatures inferred by other methods for the same time period. There is no one perfect method that can be used to measure historic temperatures over the age of the Earth. One should select a technique of determining temperature, where the data is available over the entire period of time under consideration.

Not all methods of temperature inference are of equal significance, because the technique of gathering temperature data can directly affect the precision of the results. Because of this, researchers have been known to adjust the results of collected data to make them more representable. Unfortunately, literature has several examples where these adjustments have been made, likely introducing the researcher's biases into the data. The temperature charts presented in this chapter present prehistoric inferred temperature data, but where possible, the authors have also identified the method used in temperature determination. Although no one method of gathering temperature data is possible for extended periods of time, examining the results of using several different methods can establish and confirm the trends in temperature. Verification of the trends in temperature-change, using several different methods of temperature determination is critical to understanding the Earth's climate.

Use of the Oxygen Isotope Ratio to Estimate Historic Temperatures

A critical factor defining the Earth's climatic environment is temperature. The temperature of prehistoric geological epochs can be estimated by the oxygen-isotope ratio. An example of this would be the work of Sorokhtin *et al.* (2011) who used the oxygen-isotope ratio of sea flint, which was formed in equilibrium with its surrounding ocean water. At the time of formation, the $\sigma^{18}O$ ratio in the sea flint was the same as its surrounding water as it was deposited on the ocean floor. This ratio is dependent on the temperature of the water, in which the siliceous deposits were formed. Studies of this ratio in sea flints show that for the Earth, a high atmospheric temperature existed in the Archaean time (≈3.4 BY ago) (Perry and Tan, 1972). By the mid-Archaean time oceanic water temperature rose to 72 °C (Knauth and Lowe, 1978). Other studies indicate that the oceanic water temperature for the mid-Archaean Eon (3.2 BY ago) may have reached 90 °C (Perry, 1967), and based on the isotope composition of siliceous schists even up to ≈100 °C (Knauth and Lowe, 1978). Schopf (1972) recognized the extent that $\delta^{18}O$ ratios in ancient flints reflect their origin and, the temperature of the water in which the flints had been formed. It is possible that the observed regular isotope ratio declines with age as the $\delta^{18}O$ ratios are associated not only with changes of the oceanic water temperature but also due to other reasons, e.g., with the $\delta^{18}O$ value decline in the Archaean oceanic water (Sorokhtin and Ushakov, 2002).

Sorokhtin *et al.* (2011) estimated that the seawater isotope composition in the pre-Cambrian time, as affected by the exchange reactions with iron oxides, declined to $\delta^{18}O \approx -10‰$. Galimov (1988) estimated that the Archaean oceanic water isotope composition was $\delta^{18}O \approx -15$ to $-10‰$, the $\delta^{18}O = 0$. After the disappearance of the metallic iron in the ocean from the Earth's mantle by the end of the Proterozoic time (Sorokhtin and Ushakov, 2002), probably $\delta^{18}O = 0$ for the seawater. In this case, the near-ocean-floor water temperature in Early Archaean would have been much more moderate, no higher than 70 °C. Figure 1.3 shows not only the oxygen heavy isotope $\delta^{18}O$ distribution in the seawater of various ages, but also their approximate isotopic temperature of the near-sea-floor water (considering the correction by $\delta^{18}O \approx -10‰$ as quoted above). Thus, it follows that the global ocean near-bottom-floor water temperature and the average Earth's surface temperature both began declining beginning in the Archaean time.

Historic Temperature Charts for the Past 4.6 BY

Figure 1.4, a temperature chart covering the past 65 MY, or the period that life has been on the Earth, shows not only the ocean temperature, but also the *flow of energy* (primarily from the Sun) that has warmed the surface of the Earth over time. The inferred temperatures in this graph were based on the oxygen-isotope ratios obtained from polar oceanic sediments of various ages. Significant glaciations are indicated on the graph. During most of the Earth's history, global temperatures were much warmer than they are today. This figure also indicates several earlier periods of colder temperature (glaciation). Over the last 65 MY, the Earth has experienced a steady decrease in temperature.

Glacial Periods and Interglacial Periods (4.5 to 0.54 BY AGO)

Examination of Figure 1.2 between 4.5 to 0.54 billion years (BY) ago indicates four major glacial and interglacial periods occurred in the early geologic history of the Earth. The Earth's surface, during these cooler periods, was often covered with massive sheets of ice, sometimes covering entire continents. This had a pronounced impact on: (1) Earth's climate, (2) sea levels (total oceanic water volume), (3) climatic belts, and (4) distribution of biota throughout the world (Sorokhtin *et al.*, 2007).

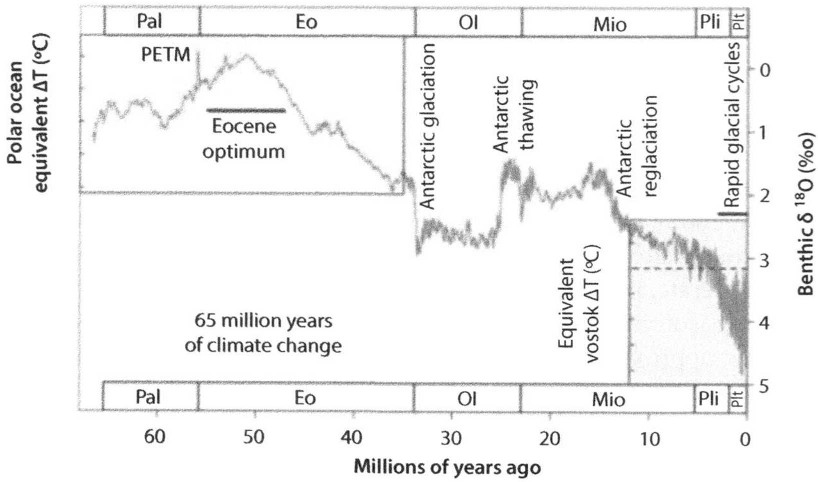

Figure 1.4 Historic deep ocean temperatures based on the 18O isotopic shift in the benthos fauna of carbonates. (isotope data from Zachos *et al.*, 2001.)

A traditional explanation for the development of these major ice ages in the past geologic periods has been a variation in the Earth's distance from the Sun, e.g., variation in the Milankovitch orbital cycles (also see http://earthobservatory.nasa.gov/Library/Giants/Milankovitch). The parameters of the Earth's orbit exhibit semi-cyclic changes with periods ranging from 0.1 to 2 MY. Sorokhtin (2005) has noted that corresponding changes in temperature at the Earth's poles can reach up to 15 °C. The periodic variations in the orbital parameters of Earth are correlated with the occurrence of glacial periods (http://geosciences.ou.edu/). This data on this temperature chart (Figure 1.4) show historic cycles of global temperatures going from warm-to-cool-to-warm with interglacial periods in between.

Historic Temperature Record of the Past 540 MY

Scotese (2003) presented the Earth's temperature chart showing relative changes in global average temperature of the past 540 MY (Figure 1.5). These temperatures of the Earth's ancient climate were computed by the Paleomar Project utilizing several methods of inferring temperature, which is reflected in the use of various time scales. This project mapped the past positions of the continents and showed the distribution of rock types that had formed in the specific time periods, recognizing the climate required to produce that rock type. Climatic conditions were inferred, as certain types of rocks, e.g., coals, require abundant rainfall and only form under specific conditions (tropical rainforests or temperate forests). Paleomar's diagrams

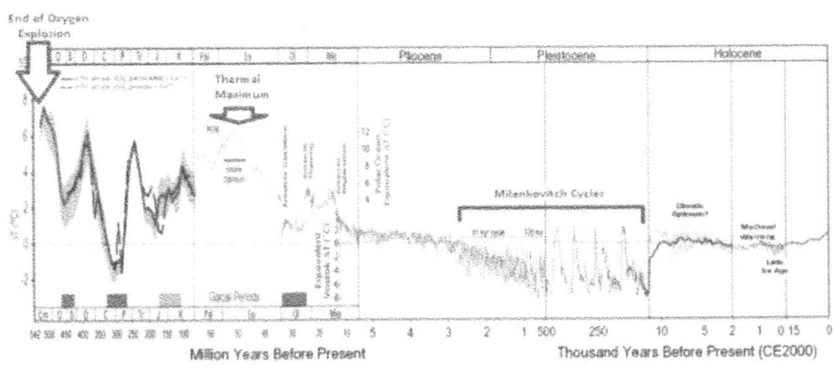

Figure 1.5 Temperature chart for the past 540 MY based upon several methods of temperature inference from various researchers. The horizontal time scale is a mixture of linear and logarithmic scales. Note that the Earth has been generally warmer in the past than it is today. (Data from U.S. DOE; in: http://theconversation.com/sudden-global-warming-55m-years-ago -was-much-like-today-35505.)

indicate which rock types formed in which climatic belts. By mapping the past distribution of thousands of these rock types, they reconstructed this chart inferring the ancient temperatures. This data indicates a shift of global temperature, alternating between frigid and warm, in the earlier periods of geologic time to almost a steady decline of temperature over the past 65 MY. Again, this temperature chart demonstrates that today's global temperatures are significantly lower than those in the past. The horizontal line in the figure represents the value of today's average global temperature.

Examination of temperature charts for the various periods of time helps identify those factors influencing the Earth's climate. Figures 1.2 and 1.5 clearly display the energy transfer from the Sun to the Earth over time. The declining temperature indicates that there is a decay in absorbed energy by the Earth occurring during the past 65 MY. Other temperature charts, using different methods of inferring temperature, support this decay of energy flow. This decline of temperature suggests a continual decrease in the Earth's absorption of energy over this period. This decay is so small that it cannot be seen if one is looking at temperature charts that cover only a few years; however, this decay of temperature can only be seen over millions of years. This absorption of energy is the primary climatic factor controlling the Earth's temperature, demonstrating that the Earth, due to decreasing absorption of energy, continues to cool today. Figures 1.2, 1.4 and 1.5, utilizing various methods of inferring the temperature, indicate a continuous decrease in the energy absorbed by the Earth with respect to time, which directly affects the Earth's climate.

These charts also show, historically, that the Earth has been cooling over the past 65 MY and continues to cool today regardless of minor cyclic variations in the Earth's temperatures. This Earth temperature chart (Figure 1.5) was assembled by the U.S. Department of Energy. Although there may be problems with the specific temperature values which were obtained by various methods, this chart clearly shows a declining trend in the Earth's temperature. What is also shown in Figures 1.2, 1.4 and 1.5, is that between 450 and 65 MY ago (C3, 2016): (1) the historic climate of the Earth had major cooler (glacial) and warmer periods lasting for long periods of time; (2) the length of time for these earlier cycles of cold and warmth far exceed what one observes in the past 65 MY; (3) the temperatures observed in today's global cooling era are neither extreme nor unique, even when compared to the relatively recent period of the Minoan/Bronze age civilization (Figure 1.6); (4) long-period temperature chart data indicates that we are in a cooling period (either glacial or interglacial) and not a warming one.

Figure 1.7 displays a cyclic relationship between the Sun and the Earth over the past ≈1 million years governed by the distance between the two

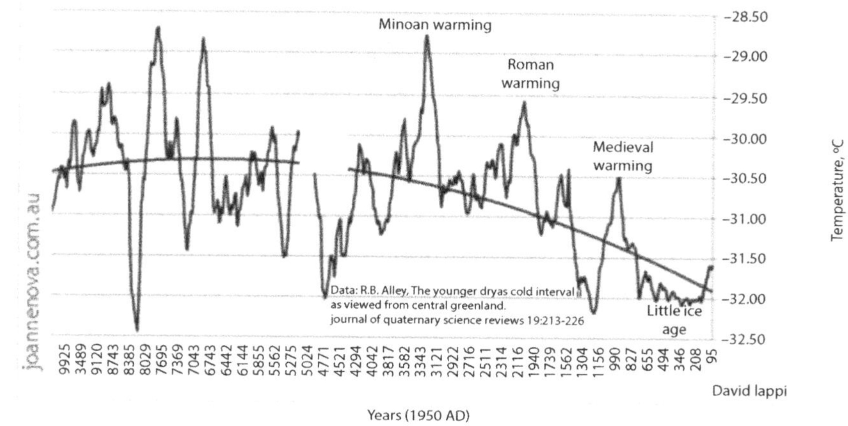

Figure 1.6 Temperature chart of Greenland interglacial temperatures for the last 10,000 years as determined from ice-sheet cores. (After Lappi, 2010.)

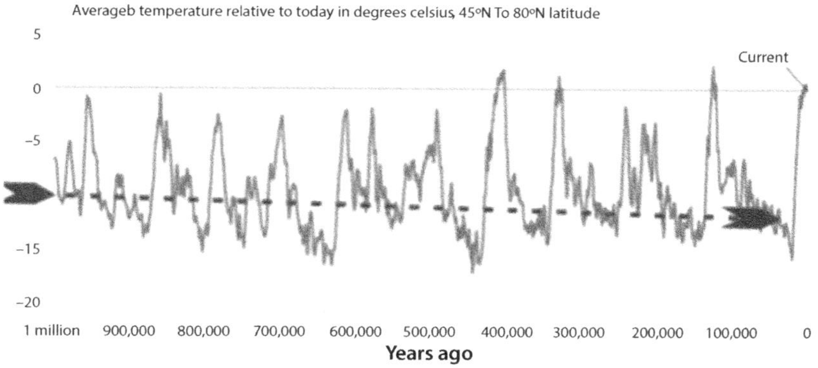

Figure 1.7 Temperature fluctuations over the past one million years. Global reconstructed temperature. (After, R. Britanja and R.S.W. van der Waal, "Global 3Ma temperature, sea level, and ice volume reconstructions." National Oceanic and Atmospheric Administration, August 14, 2008. https://www.ncdc.noaa.gov/paleo/study/11933 (Accessed April 5, 2016.)

bodies. Cycles of 100,000 years are indicated by this chart, where temperature is governed by the varying distance between the Sun and the Earth (*orbital distance*). The reason why one examines charts of global temperature over various periods of time, utilizing various methods of inferring temperature, is that charts of different periods of time often emphasize different factors that influence the Earth's climate over that time period. In this case, Figure 1.7 demonstrates that there has been a cyclical period for temperature of ≈100,000 years over the past ≈1 MY. Any temperature chart showing less than 100,000-year period would not clearly show this Earth/Sun separation-distance or *Earth's orbital effect* on the Earth's temperature. The primary controlling climatic factor of a continual cooling of the Earth was shown by charts (Figures 1.2, 1.4 and 1.5) and a secondary factor is shown over a 1 MY time period (the cyclic nature of Earth's temperature is demonstrated in Figure 1.7). Today we see people attempting to make future temperature forecasts based upon brief time periods of 25,000 or 1,000-year time periods. These periods of time are far too short to recognize the primary and secondary climate forces controlling the change of the Earth's temperatrure.

There are several problems with the horizontal scale for inferred temperatures utilizing a variety of temperature inferring methods in Figure 1.5 (C3, 2016): (1) The DOE scientists who produced this chart utilized various methods of inferred temperature data and treated all inferring methods as equal on this temperature chart, e.g., mixed methods of data determination were used (instrument thermometer values, ice core data, and dendroclimatology reconstructed estimates); (2) This chart contains swaths of historic temperature data in the chart that has had the climate records "adjusted" upward by at least 0.4 °C over the last 20 years; and (3) Much of the information in this chart was estimated by the dendroclimatology method. However, this chart is valuable as a comparison to other charts prepared utilizing different techniques of inferring temperature and identifying trends in the Earth's climate.

An example of the dendroclimatology method, used to prepare Figure 1.7, would be the use of tree rings to tree ring thickness and temperatures and then relate it to age. Generally, a year's growth of tree ring responds to changes in climatic variables by speeding up or slowing down the growth, which, in turn, is reflected in a greater or lesser thickness in growth rings. A historic tree-ring record can be established by compiling information from many living trees in a specific area. This history identifies climatic condition by year for the area. These studies can be extended to other areas by comparing the data to other groups of trees. Older intact wood that has escaped decay can extend the time covered by the record by matching the ring depth changes to contemporary specimens for several thousand years. By using this method, one can obtain tree records for one area and project the climatic data to other similar areas. Growing temperatures of older wood not connected to a contemporary record of group of trees can be dated with radiocarbon techniques and the rings then utilized to determine climatic data for temperatures of that period. A tree-ring record is also useful to produce historic information regarding precipitation, temperature, hydrology, and sometimes fire corresponding to a particular area.

Another example of the use of dendroclimatology is looking at various species (either flora or fauna) or their fossils, which respond in an analogous manner to changes in climate. There is a temperature range in which they can exist.

Geologists often utilize dendroclimatology techniques to determine climatological data for the sedimentary record (Wikipedia, 2016):

1. *Sediments*—sometimes lithified to form a rock, may contain remnants of preserved vegetation, animals, plankton or pollen, which may be characteristic of certain climatic zones.
2. *Biomarker molecules*—such as the alkenones may yield information about their temperature of formation.
3. *Chemical signatures*—particularly Ca/Mg ratio of calcite in Foraminifera tests, can be used to reconstruct past temperatures.
4. *Isotopic ratios*—can provide further information. Specifically, the $\delta^{18}O$ record responds to changes in temperature and ice volume, and the $\delta^{13}C$ record reflects a range of factors.

Underwater life, as shown in Figure 1.1, developed on Earth during the Paleozoic time (542 to 251 MY ago). This was a period that began with the breakup of the supercontinent Rodina (Pappus, 2013). The climate of this period has been determined by the dendroclimatology technique using a variety of geological evidence to deduce ancient temperatures.

In Figure 1.5, palaeotemperature charts of various geologic periods, estimated by the dendroclimatology technique, have been compressed together to show the global temperatures of the Earth over a period of 540 MY. Because of compression of several graphs, the horizontal axis (time) includes several different time scales (a mixture of log and linear). However, this figure is a useful temperature chart illustrating the exponential decay in Earth's temperature of ≈25 °F over the last 65 MY.

The temperature data presented in Figure 1.5 is based on work by Scotes (Paleomar Project) who calculated the approximate global temperatures required for the distribution of, (1) ancient corals, (2) desert deposits, (3) tropical soils, (3) salt deposits, (4) glacial material, and (5) the fossils of flora and fauna sensitive to climate, e.g., alligators, palm trees, mangrove swamps, etc. This figure illustrates that for much of this period, the average global temperatures were about 25 °C, but for short periods of time has alternated between frigid (recent age) and warmer (Cambrian, Jurassic, etc.) periods. Heib (2009) has also noted that (1) the geologic record shows that the Earth has often been much warmer and more humid than it is today; (2) the only period of the Earth's history that appears to have similar values of global temperature and atmospheric carbon content is 300 MY ago, during the late Carboniferous Period; and (3) the last 65 MY show a decrease in global temperatures of approximately 25 °C.

The late Precambrian climate, ≈600 MY ago, had several major glaciation events covering much of the Earth, resulting in a cooler period. The Cambrian Period of the Paleozoic (≈530 MY ago) was a time of explosion of organic life on the Earth. Pappus (2013) has noted that this period included the evolution of arthropods (ancestors of insects and crustaceans) and chordates (ancestors of organisms with rudimentary spinal cords). During this period, the continent of Rodina fragmented into Gondwana (which would eventually become today's continents of the Southern Hemisphere) and smaller continents made up of bits and the pieces of the land which would eventually make up today's northern continents. These changes in life and shifting of continents also affected the climate of the Earth's surface.

Another method of inferring temperatures for this time period was suggested by Zachos *et al.* (2001) using the ^{18}O isotropic ratio in benthos fauna of carbonates (see Figure 1.4). Looking at the temperatures determined from deep-water sediments, one sees a similar temperature trend when compared to the Paleomar Project data, thus verifying the global trend of a temperature decrease over the past 65 MY.

Literature is filled with global temperature forecasts ignoring these historical facts and projecting erroneous predictions of future temperatures for the Earth based only on 1,000 years or less of historical temperature data. They do not recognize their fundamental error that Earth's global climate has been cooling for over 65 MY. They also ignore the fact that ***without either a new source of energy or a change in Earth's atmosphere's capability to absorb heat, there cannot be an increase in temperature***. The ignorance of these facts has resulted in both fictitious and imaginary projections of the Earth's future global temperatures, melting glaciers, rising sea levels, and climate calamities that do not reflect reality.

Today's Temperature Charts

Today there are a vast variety of tools to measure and record the Earth's atmospheric and surface temperatures. The advent of satellites over the past few years has permitted detailed studies of atmospheric temperature from

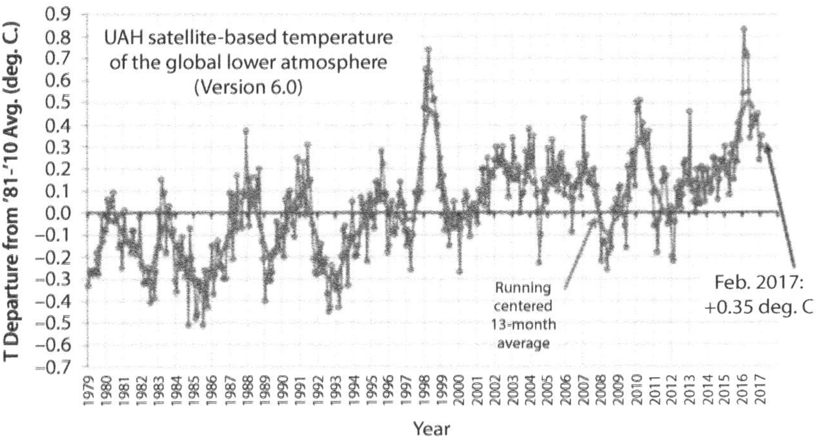

Figure 1.8 Temperature chart, 22-year history of UAH satellite-based temperature measurements of the global lower troposphere. (After Spenser, 2017.)

outside of the Earth. The temperature chart in Figure 1.8 is a 22-year chart utilizing the UAH satellite-based measurements of the Earth's atmosphere. The base value of temperature (Oct,1981) is its x-axis. Between 1979 and 2017, the value of the temperature has fluctuated, but never more than 0.9 °C. The conclusion is that over this period, there has been no significant temperature increase or decrease; however, to see a creditable change of the Earth's temperatures, as indicated by the temperature charts, requires time periods of thousands of years.

The Sun—a Primary Source of Energy

The Sun is the primary source of energy for the Earth's climate system. This concept is the key to understanding the Earth's temperature charts and energy balance between the Earth and the Sun. The Sun (1) warms the planet, (2) drives the hydrologic cycle, and (3) makes life on the Earth possible. The temperature charts (Figures 1.2, 1.4 and 1.5) demonstrate, that today's Earth is absorbing less energy from the Sun than 65 MY ago, which is reflected in the decline of the Earth's temperature.

There are three primary sources of energy to the Earth: (1) Sun, (2) radio-active material within the Earth's core, and (3) flow of energy throughout the universe. Of these sources of energy, the Sun is by far the greatest source of energy to the Earth.

Physical Aspects of the Sun

The Sun is a medium-sized star within the galaxy: its mass places it among the top 10% of its stars in the galaxy (Nine Planets, 2017). The largest object in our solar system is the Sun, which contains more than 99.85% of the total mass of our solar system. The planets contain only 0.135% of the mass of the solar system, with Jupiter containing more than twice the mass of all the planets combined. Jupiter is the second-largest member of our solar system.

The composition of the Sun is ≈70% hydrogen and ≈28% helium by mass. This composition changes slowly over time as the Sun utilizes the hydrogen as a fuel and converts it to helium. The Sun rotates around its axis like that of the Earth; however, the Sun acts as a plastic body and not a solid body like the Earth. The Sun's rate of rotation is from east to west, but its velocity varies throughout the physical sphere of the Sun. At the Sun's equator, the Sun's surface rotates once every 25.6 days, whereas near the poles it is slower (about 33.5 days). Viewed from the Earth as it orbits the Sun, the apparent rotational period of the Sun at its equator appears to be ≈28 days. This differential physical rotation of the Sun also extends down into its interior; however, the Sun's core rotates like that of a solid body (Space Weather, 2016).

Sunspots

The core of the Sun is approximately 25% of its radius. The temperature at the core is 15.6 million degrees Kelvin and the pressure is 250 billion atmospheres. At the surface of the Sun, the photosphere temperature is upward of 5,800 degrees K. The surface of the Sun (photosphere) has cooler areas, that are upward of 3,800 degrees K and appear as darker areas, known as sunspots. Sunspots vary in size, some as great as 50,000 km in diameter. *Sunspots* look dark only in comparison with the brighter and hotter regions of the Sun's surface (Nine Planets, 2017). Sunspots form where magnetic field lines come up from the Sun's interior through the solar surface causing every sunspot to have its own polarity. A sunspot consists of two parts: (1) the dark portion (umbra) and (2) the lighter part around the dark part (penumbra). The sunspot regions follow the rotation of the Sun. In size, there is a great variation; some are many times larger than the Earth and there are many that are much smaller. A large sunspot can have a temperature of 3700 °C compared to the surrounding typical photosphere temperature of about 5500 °C. The sunspot region travels across the solar disk from east to west as seen from the Earth (Space Weather, 2016).

The number of sunspots appearing in the Sun's photosphere varies cyclically, from none or a few, to a much higher number. The average length of time for this cycle appears to be about 11 years. But the length of the cycle varies and some recent investigators are suggesting a longer cycle period. Between 1700 and today, the sunspot cycle (from one $solar_{min}$ to the next $solar_{min}$) has varied in length from nine to 14 years. Figure 1.9 presents the data for the Sun's irradiance, solar flares and number of sunspots from 1975 to 2005. The term Solar maximum ($solar_{max}$) is the period of greatest number of sunspots in the solar cycle of the Sun. Around the $solar_{min}$, only few or even no sunspots can exist. As shown in Figure 1.9, the number of sunspots and the Sun's irradiance are related.

The energy produced by the Sun is $\approx 3.86 \times 10^{33}$ ergs/s or $\approx 5 \times 10^{23}$ horsepower by nuclear fusion. The Sun consumes $\approx 7.00 \times 10^{8}$ tons of hydrogen each second, converting it to $\approx 6.95 \times 10^{8}$ tons of helium and $\approx 5.00 \times 10^{8}$ tons ($3.86e^{33}$ ergs) of energy in the form of gamma rays. As energy travels through the Sun (out toward the surface of the Sun) it is continuously absorbed and re-emitted at lower temperatures (frequencies). By the time energy reaches the surface of the Sun, it is primarily a visible light.

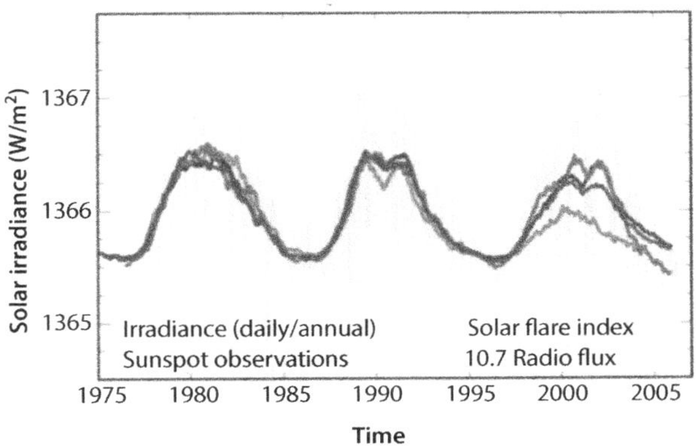

Figure 1.9 Solar cycle variation of sunspots, irradiance and solar flare index. Solar flare index is a measure of the solar radio flux per unit frequency at a wavelength of 10.7 cm. The vertical scales for each quantity have been adjusted to permit over plotting on the same vertical axis as TSI. Temporal variations of all quantities are tightly locked in phase, but the degree of correlation in amplitudes is variable to some degree. (After Wikipedia, 2017; in https://en.wikipedia.org/wiki/Solar_cycle.)

For the last 20% of the pathway to the Sun's surface, the energy is carried more by convection than by radiation (Nine Planets, 2017).

Solar Irradiation Reaching the Earth

Figure 1.10 shows the variation of the Sun's ratio,

$$\left[\frac{\text{Solar irradiation}_{\text{value at various times}}}{\text{Solar irradiation}_{\text{todays value}}}\right],$$

for luminosity, radius, and effective temperature versus time. This figure shows that with time, there has been an increase in the Sun's radius and luminosity and effective temperature.

Figure 1.11 is a temperature chart showing the annual mean surface temperatures, of the contiguous United States, between 1880 to 2006. The trend of variation of temperature over this period, calculated by the least squares technique, was an increase of 0.5 °C/century. Figure 1.12 superposes a graph showing the values for solar irradiance during the same period.

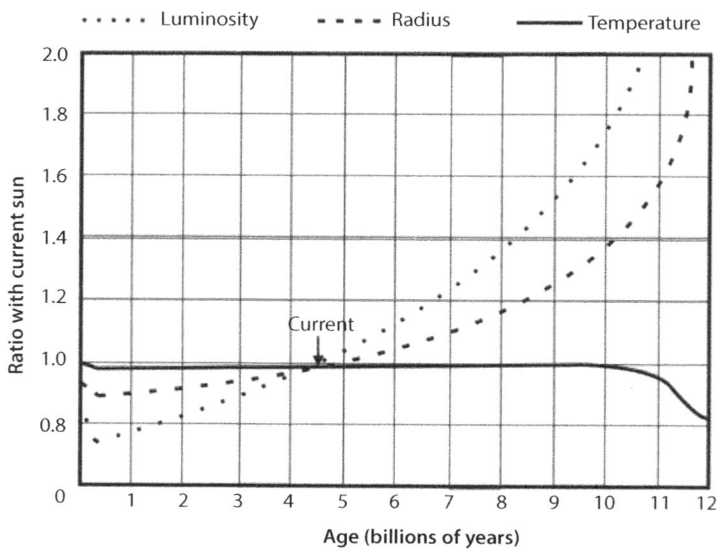

Figure 1.10 Variation of the (1) ratio of solar irradiation to the present day and (2) to the sun's luminosity; and (3) the sun's radius with time. (After Ribes, 2010.)

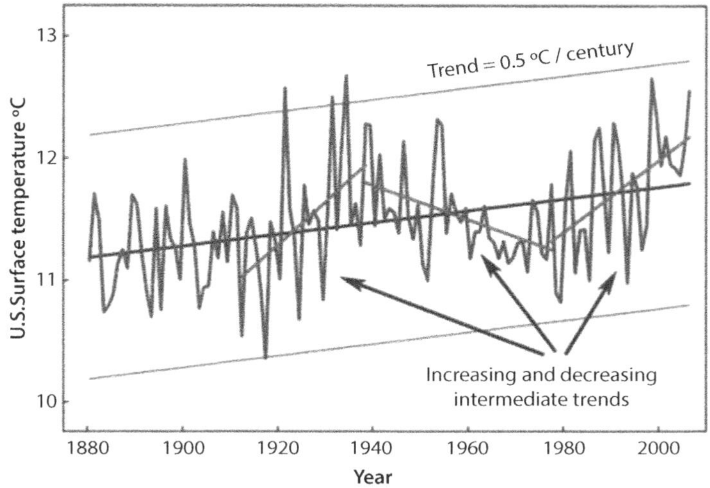

Figure 1.11 Temperature chart showing the annual mean surface temperatures in the contiguous United States between 1880 and 2006 (NCDC, 2007). The slope of the least-squares trend line for this 127-year record is 0.5 °C per century. (After Robinson *et al.*, 2007.)

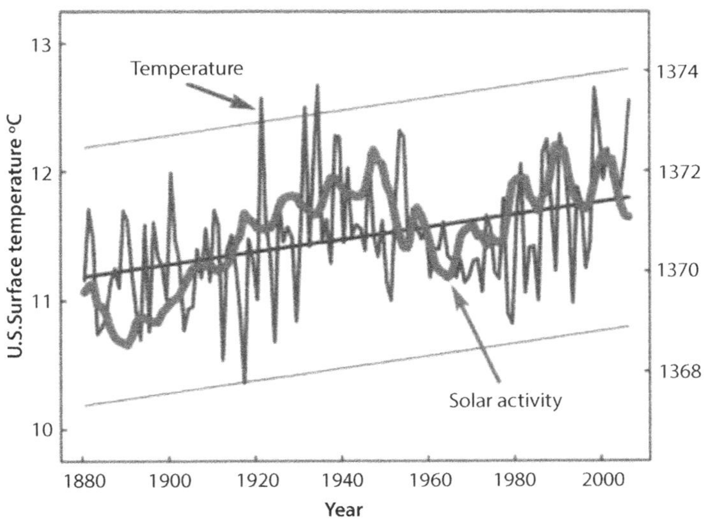

Figure 1.12 Annual mean surface temperatures in the contiguous United States between 1880 and 2006 (NCDC, 2007). The slope of the least-squares trend line for this 127-year record is 0.5 °C per century. (After Robinson *et al.*, 2007.)

Robinson *et al.* (2007) noted the percent carbon-content (carbon dioxide and methane) in the atmosphere does not correlate with the observed Earth's surface temperatures; however, solar activity correlates very well with surface temperature (see Figure 1.12). This does not exclude other potential relationships between temperature fluctuation and other phenomena; however, variations in percent carbon in the atmosphere appear to have a negligible effect on temperature chart fluctuations. Several other studies have recognized a similar correlation between surface temperature, carbon dioxide and solar activity [Baliunas and Soon (1995); Neff *et al.* (2001); Jiang *et al.* (2005); and Maasch *et al.* (2005)].

Robinson *et al.* (2007), who prepared Figure 1.11, also discussed the uncertainties introduced to temperature data by limited (short) time records and sampling locations used by researchers, e.g., if the Arctic air temperature data before 1920 had not been available for Figure 1.11, essentially no increase in temperature would have been observed for this period. A problem often occurring in temperature data is the question of what, who and why specific data was included or rejected. In this situation, Robinson *et al.* (2007) acknowledged that the inclusion or exclusion of specific temperature data would show a downward rather than an upward trend. When researchers report findings, they need to review not only the source and location of the data, but also look at long periods of time, not just a few years. Otherwise the results of their research are possibly influenced by local phenomena and/or their bias.

Variation in solar activity is not limited to just the Sun, but appears to be typical of other stars that are similar in size and age to ours (Baliunas *et al.*, 2007). It should be noted that observers have recognized warming trends rather than cooling trends on other planets and moons with different atmospheres: (1) Mars (Fenton *et al.*, 2007); (2) Jupiter (Marcus *et al.*, 2004); (3) Neptune (Hammel *et al.*, 2006; Hammel *et al.*, 2007); (4) Neptune's moon Triton (Elliot *et al.*, 1998): and (5) Pluto (Elliot *et al.*, 2003; Sicardy *et al.*, 2003). This warming trend on other planets may partially be the result of an expanding Sun (see Figure 1.10) or changes in the atmospheric pressure due to changes in the composition of the planet's atmosphere.

The variance in temperature found in the Earth's temperature charts indicates a continuous reduction in the flow of energy between the Sun and Earth for the past 65 MY. The major source of energy that the Earth receives and is retained by the Earth's surface and its atmosphere, is directly from the Sun. The absorbed energy from the Sun directly affects the Earth's temperature. There are other sources of energy, e.g., (1) energy which flows from the core of the Earth, fueled by the breakdown of radioactive materials within the Earth's core; and (2) energy that flows from the center of the

Table 1.1 Sources of Earth's heat flow - First order climate drivers. (After Sorokhtin *et al.*, 2007.)

Source of heat.	**Estimated heat**	**Percent of total energy**
Directly from Sun reaching Earth's surface	17500×10^{20} ergs/s	99.95%
From the Earth's core transmitted through crust	4.3×10^{20} ergs/s	0.025%
Generated by man	1.34×10^{20} ergs/s	0.0077%

universe and passes through our solar system. As discussed in the later chapters, as the constituents of the atmosphere vary, the atmospheric pressure varies and so does the ability of the Earth's atmosphere to absorb heat. The total average solar energy flux currently reaching today's Earth's surface is $S_0 \approx 1.75 \times 10^{24}$ erg/s, which is determined by the so-called solar constant $S \approx 1.37$ kW/m^2 or 1.37×10^6 erg/cm^2 s (Horrell, 2003). The total heat flux through the Earth's surface due to energy generated within the Earth's core and transmitted through the mantle and crust is estimated at $\approx 4.3 \times 10^{20}$ erg/s (Sorokhtin and Ushakov, 2002), which is ≈0.0257% of the total Earth's solar irradiation. The world's total energy production in the year 2003 was $\approx 1.34 \times 10^{20}$ erg/s (Key World Energy Statistics, 2004), which is ≈0.0077% of the total solar irradiation reaching the Earth.

Table 1.1 shows that the solar irradiation is the dominant source of energy to the Earth's atmosphere, continental surface and hydrosphere. One can estimate that solar irradiation supplies more than 99.95% of total energy driving Earth's climate. Thus, heating and cooling of the atmosphere is primarily due to variations in insolation of the Earth (Hoyt and Schaten, 1997). Comprehensive data on the Sun as a primary source of energy for the Earth's climate are presented by Pekarek (2001).

The Sun's Energy

The Sun is ≈4.65 BY old. Since its birth, it has used up approximately half of the original hydrogen stored within its core as fuel (Nine Planets, 2017). The Sun has sufficient fuel to continue to radiate energy *peacefully* for another ≈5 BY, when it will run out of hydrogen fuel. The Sun will then destroy itself and its associated solar system. Its luminosity has gradually increased throughout the time of its existence and the present-day energy output is about 40%

higher than that at the time of its formation (Gribbin, 1991). This energy output is expected to increase by about 15% in the future 1.5 BY. Figure 1.10 illustrates Ribes (2010) projection for the increase in the Sun's radius, luminosity and effective temperature with respect to time. The significant fact is that the Sun's energy, size and temperature are increasing with time.

The Sun radiates electromagnetic energy throughout the entire electromagnetic spectrum. The Earth's atmosphere absorbs and reflects this radiation from the Sun (Figure 1.13). Today's satellite technology provides the measurements that one needs to measure the Total Solar Irradiance (TSI) for the distance from the Earth to the Sun. These satellite observations are required to develop a quantitative climatic theory and explanation of changes in the Earth's climate. The satellite data measurements have been available since 1978 (Wilson, 1997). In lieu of direct satellite measurements, one can also use variables correlated with TSI, e.g., the number of sunspots recorded since 1610, or solar magnetic activity (Baliunas and Saar, 1992; Baliunas and Soon, 1996). The Total Solar Irradiance (TSI) is cyclic, dependent upon the number and size of the sunspots, being strongest at the peak of magnetic activity (maximum number of sunspots) and varying directly with the intensity of Sun's activity. Tinsley (1997) has noted that the statistical confidence for the correlations between the number of sunspots and irradiation is high.

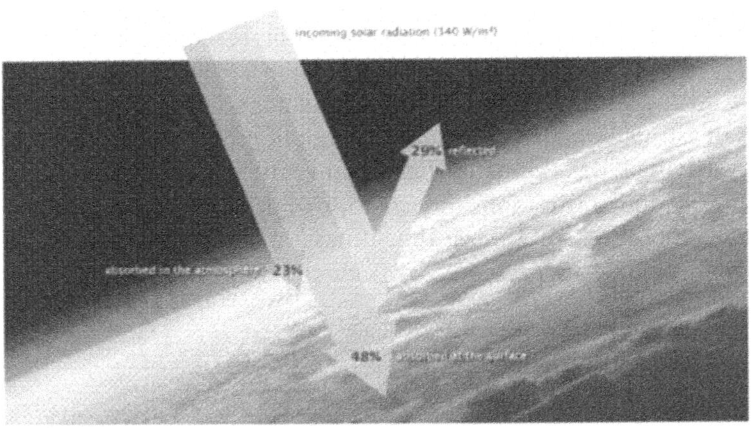

Figure 1.13 Absorption of sun's energy by the Earth's atmosphere. Of the 340 watts per square meter of solar energy that falls on the Earth, 29% is reflected into space, primarily by clouds, but also by other bright surfaces and the atmosphere itself. About 23% of incoming energy is absorbed in the atmosphere by atmospheric gases, dust, and other particles. The remaining 48% is absorbed at the surface. (After NASA illustration by Robert Simmon. Astronaut photographISS013-E-8948.)

As noted earlier, the energy output of the Sun is not constant and is related to the Sun's size (Figure 1.10) and number of sunspots, which vary daily over an ≈11-year cycle. A historic period of very low sunspot activity occurred in the latter half of the seventeenth century, referred to as the *Maunder Minimum*. It coincided with an abnormally cold period in northern Europe, often called the *Little Ice Age* (see Figure 1.6). Since the formation of the solar system, the Sun's luminosity has increased by about 40% (Figures 1.10 and 1.14).

The effect of solar irradiation on global atmospheric temperature can also be evaluated using the adiabatic model of heat transfer within the Earth's atmosphere. Analysis of the Sun's temperature changes, attributed to variations in energy and matter flux, can be made using the following formula:

$$\frac{T_h}{T_o} = \left(\frac{S}{S_o}\right)^{\frac{1}{4}} \left(\frac{p_h}{p_o}\right)^{\alpha} \quad \text{(Eq. 1.1)}$$

where T_h is the global atmospheric temperature at any given altitude *h*; T_o is the present-day global temperature at sea level, in degrees' Kelvin (T_0 = 288 K); *S* is the total solar energy flux reaching the Earth's surface; S_0 is the total present average solar energy flux ($S_0 = 1.75 \times 10^{24}$ erg/s); p_h is the global atmospheric pressure at the altitude *h*; and p_0 is the global average atmospheric pressure at sea level (p_0 = 1 atm).

A rough estimate of global atmospheric temperature change at sea level attributed to variations in the Sun's insolation *S*, Eq. 1.1 can be rewritten in the following form:

$$\Delta T = T - T_o = T_o\left[\left(\frac{S}{S_o}\right)^{\frac{1}{4}} - 1\right] = 288\left[\left(\frac{S}{S_o}\right)^{\frac{1}{4}} - 1\right] \quad \text{(Eq. 1.2)}$$

Global temperature changes attributed to variations in the Earth's insolation with corresponding changes in the Earth's global temperature at sea level are presented in Table 1.2. As shown in this table, a 1% increase in current solar radiation reaching the Earth's body translates directly into ≈0.86 °C increase in the Earth's global temperature. This equation can also be utilized to determine an upper estimate for a possible atmospheric temperature increase due to anthropogenic activities.

Even if the entire world energy generated by humans, estimated at $\approx 1.34 \times 10^{20}$ erg/s, could be utilized for only heating the Earth's atmosphere,

Table 1.2 Global Temperature Changes Attributed to Variations in Earth's Insolation. (After Sorokhtin *et al.*, 2007.)

S/So	0.85	0.90	0.95	0.99	1.00	1.01	1.05	1.10	1.15
AT, °C	−11.5	−7.49	−3.77	−0.86	0.00	0.86	3.77	7.49	11.5

the corresponding atmospheric temperature increase would not exceed 0.01 °C at sea level. If, in addition, one takes into consideration that changes in the global atmospheric temperature are closely correlated with the changes in solar activity as indicated in Figure 1.9, then the conclusion is that solar irradiation must be the dominant source of energy driving the Earth's climate (see also Kondratiev, 1992; Hoyt and Schatten, 1997). Later chapters will discuss the effect of changing the composition of the Earth's atmospheric and its ability to absorb energy.

As shown in Figure 1.13, about 29% of the solar energy that arrives at the top of the Earth's atmosphere is reflected into space as: (1) clouds, (2) atmospheric particles, (3) bright ground surfaces such as sea ice and snow and (4) so forth. This reflected energy plays no role in the Earth's climate system. About 23% of the incoming solar energy is absorbed in the atmosphere by water vapor, dust, and ozone, and 48% passes through the atmosphere and is absorbed by the Earth's surface. Thus, about 71% of the total incoming solar energy is absorbed by the Earth system (NASA, 2017).

Energy Received by the Earth from the Sun

Reviewing the short-term temperature chart (Figure 1.11), which presents the history of the annual mean surface air temperature of the contiguous United States between 1880 and 2006, Robinson *et al.* (2007) determined the slope of this temperature chart (Trend ≈0.5 °C/Century) using a least-squares method. This natural warming trend agrees with other surface temperatures observed for the United States during the past century, even though this data may be suspect due to the selection process of what temperature data should be included or excluded in its base data. However, Robinson *et al.* (2007) noted that their compiled U.S. average surface temperatures show an increase of ≈0.5 °C per century, which is consistent with other historical rate values found in literature of 0.4 to 0.5 °C per century during the similar time period from the Little Ice Age, as determined by Groveman and Landsburg (1979), Esper *et al.* (2002), Tan *et al.* (2004) and Newton *et al.* (2006).

Again, as noted earlier, the Earth's atmospheric temperature is primarily regulated by the Sun, which fluctuates in solar activity as measured by (1) sunspot cycle amplitude, (2) sunspot cycle length, (3) solar equatorial rotation rate, (4) fraction of penumbral spots, (5) variations in Earth's atmospheric composition, and (6) decay rate of the 11-year sunspot cycle (Soon, 2005, and Hoyt *et al.*, 1993). Between 1900 and 2000, on absolute scales of solar irradiance and degrees Kelvin, solar activity increased 0.19%, whereas the temperature change was 0.21% (Robinson *et al.*, 2007). It is estimated that the Earth's temperature would be reduced by 0.6 °C by particulate blocking of the Sun's radiation or 0.2% (Teller *et al.*, 1997). Figure 1.12 demonstrates that there is a close relationship between the solar activity and the U.S. average surface temperatures. The total solar irradiance can be measured by the: sunspot cycle amplitude, sunspot cycle length, solar equatorial rotation rate, fraction of penumbral spots, and decay rate of the 11-year sunspot cycle (Soon, 2005; Hoyt and Schatten, 1993).

In summary, the solar irradiance correlates well with Arctic temperature charts, whereas the content of carbon in the atmosphere does not (Marland, 2007). Robinson *et al.* (2007) have noted, that for particularly short periods of time, e.g., 100 years, uncertainties can be introduced into the temperature charts by limited time records and the location of temperature sampling. For example, the Earth's temperature charts in Figures 1.11 and 1.12 would look quite different if the Arctic air temperatures before 1920 were not included. In this case, the basic temperature chart would be affected and one would not see an increase in temperature.

The Paradox Reviewed

The climatic paradox: "***The Sun's luminosity is increasing, whereas the Earth's temperatures are decreasing***" (Figures 1.4 vs. 1.10).

The conclusions of the chapter defining this paradox are as follows:

1. Temperature charts for the sediments of the Earth's deep oceans show that temperatures have steadily declined ≈12 °C over the past 65 MY (Figure 1.4).
2. Other long-term temperature charts, e.g., Figures 1.2, 1.4, and 1.5) also show that the Earth has been cooling over the last 65 MY.
3. Shorter-term temperature charts of ≈1 MY show that Earth's temperature has been cyclic over this period along with the long-term cooling shown in the 65 MY charts (Figure 1.7).

4. The Sun is the primary source of energy warming the Earth's surface and its atmosphere, as reflected in Figure 1.12, illustrating a strong relationship between the Sun's irradiance and the Earth's temperature.
5. In preparation of temperature charts, to determine the Earth's cooling or warming, one must use long-term temperature charts greater than 0.1 MY. Shorter-term charts do not cover a sufficient period to observe the change in temperature.
6. Short-term charts of 1,000 years or less can be strongly affected by localized phenomena, e.g., the location and number of sampling as shown in Figures 1.11 and 1.12.
7. The irradiation of the Sun has increased since the formation of the Earth and will continue to increase with time (Figure 1.10).
8. The Sun's radiation, the Earth's atmosphere (composition and pressure) and the Earth's orbital distance from the Sun have not remained constant over time, resulting in a variation in absorbed energy.
9. There have been changes in the Earth's atmospheric pressure (due to a variation of atmospheric composition), which directly affect the ability of the Earth's atmosphere to absorb energy.

The original question arises: "**How *can the Sun be transmitting more energy to the Earth, and yet the Earth has been definitely cooling over the past 65 million years, particularly when the Sun is the primary source of energy to the Earth?*"**

The answer is that the ability of the Earth and its atmosphere to absorb heat is constantly changing. The reasons why are covered in the following chapters.

2

Adiabatic Theory

A process that occurs without loss or gain of heat may be defined as adiabatic. In a flow path, $1 \rightarrow 2 \rightarrow 3$, $p_1 v_1^{\alpha} = p_2 v_2^{\alpha} = p_3 v_3^{\alpha}$, where p_i is the absolute pressure; v_i is the specific volume (volume/unit weight); and α is the adiabatic exponent. Because climate (weather) occurs within the troposphere, a climatic model is essentially one of the troposphere.

Troposphere

The troposphere is the lowest subdivision of Earth's atmosphere, where nearly all weather conditions (climate) take place. Approximately 75% of Earth's atmospheric mass and 99% of water vapor and aerosol mass lies within this layer. The thickness of this layer varies: (1) at the equator (20 km or 12 mi), (2) in the mid latitudes (17 km or 11 mi), and (3) at the polar regions in the polar regions in winter (7 km or 4.3 mi). The lowest portion of the troposphere, where the maximum friction between the Earth's surface and atmosphere occurs, is known as the planetary boundary layer. This layer varies from a few hundred meters to 2 km (1.2 mi) deep depending

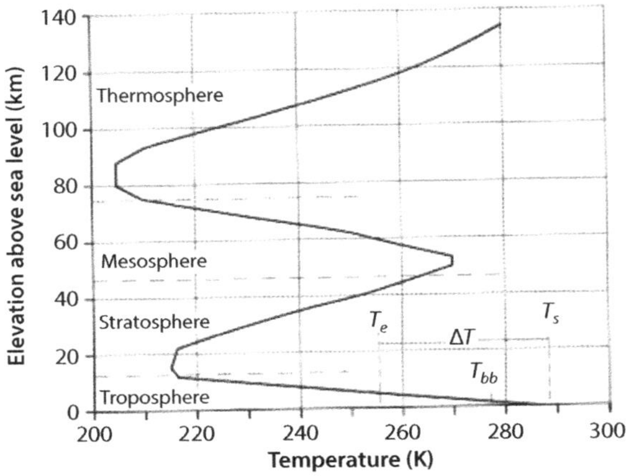

Figure 2.1 Temperature distribution vs. elevation (altitude) in the Earth's atmosphere. T_e is the Earth's effective temperature; T_s is the average Earth's temperature normalized for sea level; ΔT is the greenhouse effect value ($T_s - T_e$); T_{bb} is the absolute black body temperature at the Earth from Sun distance. (After Atmosphere of Earth, 1988.)

on the landform and time of day. The upper level of the troposphere is known as the tropopause, which is the border between the troposphere and stratosphere. This layer is a temperature inversion layer, where the air temperature ceases to decrease with height and remains constant through its thickness (Figure 2.1).

The Earth's weather occurs in the troposphere. Climate reflects the Earth's weather over a period of time. A model of the Earth's climate is essentially a model of the troposphere. The troposphere has the following characteristics (see Figures 2.1 and 2.2):

1. The boundaries of the troposphere are the Earth's surface as the lower surface and an upper surface ≈12 km (7 miles) above the Earth's surface (see Figure 2.1).
2. The troposphere is unique in that the movement of air is parallel to rather than vertical to the surface of the Earth as in the case of all other positions (layers) of Earth's atmosphere.
3. The troposphere is an exclusive layer as there is little mixing of air between the troposphere and the upper portions of the atmosphere.
4. What is often referred to as the Earth's *weather* (short time period) or *climate* (longer time period) occurs within the troposphere.

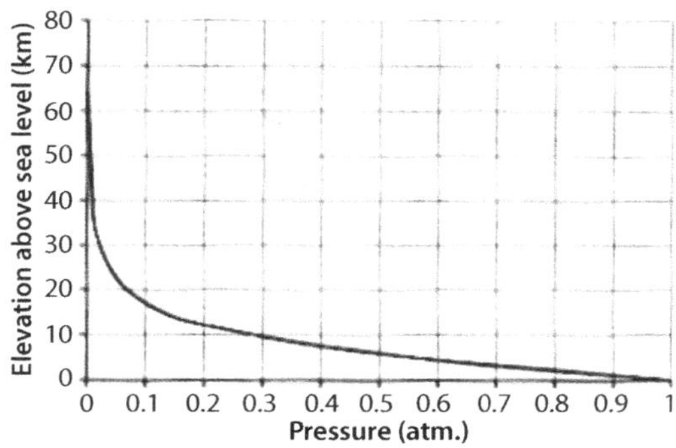

Figure 2.2 Atmospheric pressure vs. elevation (calculated using Eq. 2.1 and from the temperature vs. elevation (altitude) in Figure 2.1.).

5. Approximately 80% of the mass of the atmosphere lies within the troposphere.
6. The troposphere pressure ranges from 1000 to 200 millibars (29.92 in. to 5.92 in. or 1 atm to ≈0.3 atm)).
7. The temperature generally decreases linearly with increasing height up to the top of the troposphere (see Figure 2.2).
8. The temperature averages ≈15 °C (59 °F) near the Earth's surface and ≈ −57 °C (−71 °F) at its upper limit.
9. Approximately 90% of heat (energy) transfer within the troposphere is by convection.
10. The troposphere ends (layer) at the point where temperature no longer linearly varies with height.
11. Wind velocity increases with increasing elevation up to the jet stream.
12. The moisture concentration decreases with increasing height.
13. The Sun's heat that warms the Earth's surface is transported upwards largely by convection and is mixed by updrafts and downdrafts.

How is Heat Transferred in the Troposphere?

The absorption of energy (heat) from the Sun and its transference within the Earth's troposphere occurs by the three different processes as illustrated in Figure 2.3:

1. *Radiation* is the transfer of heat (energy) without the involvement of a physical substance in the transmission. Radiation

Figure 2.3 Various methods of energy transfer in the troposphere. (After UCAR, 2018, https://www.COVCAR.edu/learn/1_1_1.htm.)

can transmit heat through a vacuum. Energy travels from the Sun to the Earth by means of electromagnetic waves. The shorter the wavelength, the higher the quantity of energy associated with it.

2. *Convection* is the transfer of heat by transporting groups of molecules from place to place within a substance. In the atmosphere, convection includes the large- and small-scale rising and sinking of air masses and smaller air parcels. These vertical motions of air effectively distribute heat and moisture throughout the atmospheric column and contribute to cloud and storm development (where rising motion occurs) and dissipation (where sinking motion occurs). Meanwhile, the slow rotation of the Earth toward the east causes the air to be deflected toward the right in the Northern Hemisphere and toward the left in the Southern Hemisphere. This deflection of the wind by the Earth's rotation is known as the Coriolis effect.
3. *Conduction* is the transfer of heat (energy) through contact with the neighboring molecules. Air and water are relatively poor conductors of energy. Since air is a poor conductor, most energy transfer is by conduction occurring at the Earth's surface.

The origin of almost all of Earth's energy comes from the Sun. Therefore, any climatic model developed, must include: (1) the energy coming from the Sun to the Earth, (2) energy intercepted by the atmosphere (absorption and reflection) and (3) energy absorbed and reemitted by the Earth's surface.

Modeling the Earth's Troposphere

The Earth's atmosphere is an example of the dissipative system (dissipating energy), as described by the nonlinear equations of mathematical physics. Therefore, the Earth's atmosphere can be organized and described by its physical fields and formation using the process parameters-defined stable thermodynamic structures in space and time. For instance, it is possible to examine only those physical parameters, such as, the mass of the atmosphere, its heat capacity, the average value of the solar radiation hitting the Earth as well as a strong negative feedback between the planet's spherical albedo and its averaged near-surface temperature. With this set of parameters, the local details in the *greenhouse effect* description is the first approximation model, single-dimensioned and averaged over the entire Earth.

This approach has several advantages. It enables obtaining analytical and unique results in the solution of global problems as well as the influence of the atmosphere's composition on the total value of the *greenhouse effect* for the entire Earth. It is also possible to include within this single-dimension model, additional and local parameters, e.g., the latitude of a locality, the Earth's revolution axis angle with the ecliptics, its precession, inflow of additional heat by the air flows (cyclones), the snow-cover albedo, etc. It is possible to construct either a 3-D or even 4-D (with the time as the fourth-dimension) *greenhouse effect* model.

Four primary factors responsible for the climatic conditions on the Earth are: (1) the amount of solar radiation, (2) the composition and pressure of the atmosphere, (3) the heat capacity of the Earth's atmosphere, and (4) the Earth's precession angle (Sorokhtin, 1990, 2001,2006; Khilyuk and Chilingar, 2003, 2004; Sorokhtin *et al.*, 2007a, b); and Chilingar *et al.*, 2014).

Features of the Earth's Atmosphere

The total mass of the Earth's atmosphere is $\approx 5.15 \times 10^{21}$ grams. The average atmospheric pressure at sea level, p_0, is equal to one physical atmosphere or 1.0132 bars (1013.2 millibars) or 760 mm mercury. At sea level the Earth's atmospheric density, r_0, is $\approx 1.27 \times 10^{-3}$ g/cm^3. The air pressure and density rapidly decline with altitude, or height above sea level, under the exponential rule expressed below or as shown in Figures 2.1 and 2.2:

$$p = p_o \ e^{\left(\frac{g\mu}{RT}h\right)}, \qquad \text{(Eq. 2.1)}$$

where $g = 981$ cm/s^2 is the gravitational acceleration; μ is the average molecular weight of the atmospheric gas (for Earth's atmosphere at $p = p_o$, $\mu = 28.97$); R = 1.987 cal/deg · mol or 8.314×10^7 erg/deg · mol is the gas constant; T is the absolute temperature in degrees Kelvin; and h (cm) is the elevation above the sea level.

The atmospheric nitrogen-oxygen dry-air composition is composed (by mass) of: 75.51% nitrogen, 23.15% oxygen, 1.28% argon, 0.046% carbon dioxide, 0.00125% neon and $\approx$0.0007% other gases. An important active greenhouse gas component of the atmosphere is the water vapor and the drops of water found in clouds. The total water vapor or water content in the atmosphere reaches the quantity of 0.12 to 0.13×10^{20} g. This is equivalent to an average water layer, 25 mm thick, covering the Earth (the pressure equivalent of 2.5 g/cm^2). If we take the average annual humidity evaporation and the precipitation, at a pressure of $\approx$780 mm of mercury, then the water vapor is renewed within the troposphere about 30 times per year (i.e., every 12 days).

In the stratosphere, a layer of ozone molecules absorbs a sizable portion of the UV solar radiation. The heat release in the formation of ozone molecules results in heating of the air masses at elevations of about 50 km. This can be seen by the temperature profile shown in Figure 2.2. The ozone (O_3) content in the atmosphere is $\approx 3.1 \times 10^{15}$ g, whereas the oxygen (O_2) content is 1.192×10^{21} g. Ozone is mainly found in two layers of the Earth's atmosphere. About 98% of the ozone resides in layers within the stratosphere ranging from 6 to 10 miles (20 to 17 km) above sea level and extends to 30 miles (50 km).

There are several transitional layers between the troposphere and stratosphere, and between the mesosphere and thermosphere. These layers are, respectively, the troposphere, with temperatures of 190 to 220 K, and the mesosphere with temperatures of 180 to 190 K.

The thermosphere is located above the mesosphere where the ionized gas glows at a temperature of 1000 K and greater. At elevations over 1000 km the thermosphere gradually transforms into the exosphere and above this, into the outer space.

There are several significant differences within the atmosphere between the troposphere and those layers above it. In the layers of atmosphere above the troposphere, air flows vertically, away from the Earth, rather than horizontally as in the troposphere where it tends to flow in spirals and parallel to the Earth's surface. One reason for this parallel-spiral flow pattern is the friction between the Earth's surface and atmosphere; as the Earth spins on its axis, it pulls the atmosphere along with it. Another reason is the *Coriolis effect*, a result of the rotation of the Earth. The daily turbulence of this zone of air flow in the troposphere is what is referred to as weather, or on a long-term basis, climate. Due to the difference in air flow in the troposphere as compared to the upper portions of the atmosphere, the average temperature distribution in the Earth's troposphere is radically different from its distribution in the stratosphere, mesosphere, and thermosphere. Temperature distribution is almost linear in the troposphere, whereas, as shown in Figure 2.2, in the upper layers of the atmosphere it is very nonlinear, with a characteristic maximum at the elevations of about 50 km and the temperature growth above 90 km above the sea level. The temperature in the stratosphere and mesosphere is mostly determined by the heat release in the oxygen dissociation process and the radiation mechanism of heat transfer, whereas the temperature distribution in the troposphere is determined by different processes. The most important of these processes is the convective heat release from the denser tropospheric layer into the stratosphere, where it then is lost as it radiates out toward the outer space.

Development of an Adiabatic Equation

Premise of the Adiabatic Theory

The Earth's troposphere is a relatively dense atmosphere with pressures ranging from 0.3 to 1 atm. For this reason, energy (heat) transfer in this lowermost layer of the atmosphere occurs mostly due to air-mass convection (Sorokhtin, 2001) and not just to radiation as often believed by the proponents of the classic *greenhouse effect*. A dense atmosphere, with an air pressure greater than $p > 0.2$ atm, will be dominated by the heat transfer in airflows by the mechanism of convection of air mass exchange. In this process of air-mass exchange, warmer masses of air expand and rise whereas the cooler air-masses contract and descend. Radiation heat exchange is dominant only in the higher layers of the atmosphere, e.g., stratosphere, mesosphere and thermosphere layers. Thus, the average temperature distribution within the troposphere, T, must be close to an adiabatic distribution:

$$T = Cp^{\alpha}, \qquad \text{(Eq. 2.2)}$$

where, α is the adiabatic exponent, dependent only on the heat absorption capacity of the mixture of atmospheric gases. This does not mean that any specific temperature distribution, at a specific moment in time, must necessarily be entirely adiabatic. What is meant is that the average distribution, within time intervals, e.g., over a period of time, is adiabatic.

The Earth's illumination (by a flow of energy from the Sun) can be described by the temperature of an *absolute-black-body*, T_{bb}. This temperature may be determined by the Boltzmann's equation:

$$T_{bb} = \left(\frac{S}{4\sigma}\right)^{\frac{1}{4}}, \qquad \text{(Eq. 2.3)}$$

where $S = 1.367 \times 10^6$ erg/cm$^2 \cdot$ s representing the solar constant or the incoming flow of solar energy; the Stefan-Boltzmann constant is $\sigma = 5.67 \times 10^{-5}$ erg/cm$^2 \cdot$ s $\cdot$ deg. Part of the solar radiation, at a radiation temperature, T_r, is reflected from the Earth and is lost to the outer space:

$$T_r = \left(\frac{SA}{4\sigma}\right)^{\frac{1}{4}}, \qquad \text{(Eq. 2.4)}$$

where A is the planet's reflectivity or albedo (for the Earth, $A = 0.3$). In this case at $\psi = 0$, the Earth's radiation temperature, $T_r = 206$ K, where only part of the solar energy reaches the Earth, it has an effective temperature, T_e:

$$T_e = \left(\frac{S(1-A)}{4\sigma}\right)^{\frac{1}{4}} \qquad \text{(Eq. 2.5)}$$

For Earth, $T_e \approx 255$K.

The temperature determined by Eq. 2.4 is sometimes called the radiation temperature, T_r, because it describes a planet's radiation as it is seen from the outer space (although that is not totally accurate).

The greenhouse effect, ΔT, is the difference between the planet's average surface temperature, T_s; and its effective temperature, T_e:

$$\Delta T = T_s - T_e. \tag{Eq. 2.6}$$

The average temperature of the entire Earth is approximately 288 K or 15 °C, and its effective temperature according to Eq. 2.5, at ψ; = 0, is T_e = 255 K or −18 °C. Thus, the classical greenhouse effect value for the Earth is +33 °C.

Development of the Adiabatic Equation

In the troposphere, where the atmospheric pressure is more than 0.2 atm, the dominating type of heat transfer is by mass air flow through a convective mass exchange where the warmer air expands and rises and the colder air contracts and drops (Figure 2.3). Radiation heat transfer dominates only in the upper layers of the atmosphere, above the troposphere, where there is lower pressure ($p > 0.1$ atm) and where the molecules are further apart, e.g., in the stratosphere, mesosphere, and thermosphere. This suggests that the average temperature distribution in the troposphere must be close to adiabatic distribution. If no or little outside energy is added to the troposphere, then following the gas law, little energy either enters or leaves the troposphere:

$$\frac{p_i V_i}{T_i} = \text{constant}, \tag{Eq. 2.7}$$

where there is no change in the volume of the gas, the adiabatic temperature distribution is controlled by the atmospheric pressure, p and by the effective heat capacity of the air (Landau and Lifshitz, 1979):

$$T^{\gamma} p^{(1-\gamma)} = \text{consant}, \tag{Eq. 2.8}$$

and

$$\gamma = \left(\frac{c_p}{c_v}\right), \tag{Eq. 2.9}$$

where c_p and c_v are the gas heat capacities respectively at constant pressure and constant volume. T and p may be determined by the formula:

$$T = Cp^{\alpha}, \tag{Eq. 2.2}$$

where

$$\alpha = \left(\frac{\gamma - 1}{\gamma}\right) \text{ and } \left(\gamma = \frac{c_p}{c_v}\right) \qquad \text{(Eq. 2.10)}$$

Eq. 2.10 shows that under the adiabatic process, the gas temperature, T, is always proportionate with the gas pressure, p, to a power of the adiabatic exponent, α, which depends on the effective heat capacity of the gas mixture, atm. For all triatomic gases, e.g., CO_2 and H_2O, $\gamma = 1.3$ and $a_o = 0.2308$, and for the biatomic gases, e.g., N_2 and O_2, $\gamma = 1.4$. For dry air, without consideration of absorption of the Earth's heat (IR) radiation, $\alpha = 0.2857$.

Water vapor in the atmosphere generates the cloud cover which is a primary factor in determining the Earth's reflectance, A. Earth's reflectivity (albedo), A. The albedo creates a strong negative feedback between the near-Earth temperature, T_s, and the *absolutely-black-body* temperature, T_{bb}, which describes the solar irradiation intensity of Earth, S, at its distance from the Sun. Any increase in near-Earth temperature increases both the evaporation of moisture on the Earth which also adjusts Earth's cloud cover. Changes in the cloud cover will alter the Earth's albedo and reflectivity of the atmosphere. The result is a change in the reflection of the Sun's energy by atmospheric clouds to outer space and thus a change in the flow of energy to the Earth's atmosphere and surface. Because of this, the average temperature of the Earth's surface declines to its previous level. Any deep negative feedback in this system results in a linear correlation between the reaction of the system's output and the action at its input. This property of a system, with a negative feedback, is universal and manifests itself regardless of the system's nature, whether it is a planet's atmosphere, electron amplifier or Watt's centrifugal control in a steam engine.

For this reason, the near-surface temperature is always proportional to the effective temperature value, T_e (Figure 2.4).

The following conclusion can be made from this under the classic version of the theory, the near-surface temperature, T_s, and the temperature, T, in general at any level within the troposphere, are proportionate to the effective temperature of a revolving body, T_e, at the distance from the Earth to the Sun. The effective temperature, T_e depends on the solar radiation intensity, solar constant (S), and is proportional to the atmospheric pressure, p, raised to the power of adiabatic exponent, α. The adiabatic exponent depends on the composition of the atmosphere and its heat absorbing capacity:

$$T = b^{\alpha} T_e \left(\frac{p}{p_o}\right)^{\alpha} = \left(\frac{S(1-A)}{4\sigma}\right)^{\frac{1}{4}} \left(\frac{p}{p_o}\right)^{\alpha}, \qquad \text{(Eq. 2.11)}$$

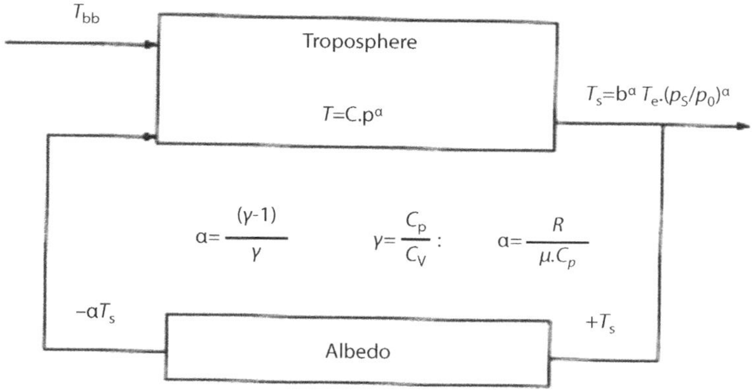

Figure 2.4 Flow diagram of the temperature transformation in the troposphere. T_{bb} is the absolutely black body temperature, K, at the distance of the Earth from the Sun; T_e is the effective Earth's temperature, K; T_s is the near-Earth temperature, K; p_s is the atmospheric pressure; p_o pressure unit; b is the scaling factor; α is the adiabatic exponent; c_p is the air heat absorbing capacity at constant pressure; c_v is the air heat capacity at constant volume; R is the gas constant; m is the molar weight of the air gas mixture.

where, p is the troposphere's atmospheric pressure; p_o is the pressure unit, for instance, $p_o = 1$ atm; α is the adiabatic exponent:

$$\alpha = \frac{(\gamma - 1)}{\gamma}, \quad \text{(Eq. 2.10)}$$

and

$$\gamma = \frac{c_p}{c_v}, \quad \text{(Eq. 2.9)}$$

where c_p and c_v, represent the specific heat of the air mixture at constant pressure and volume, respectively; b is the scale coefficient presented in Eq. 2.11 is equal to as:

$$b = \left[\frac{1}{(1-A)^{\frac{1}{4}}}\right]^{\frac{1}{\alpha}} \quad \text{(Eq. 2.12)}$$

If the specific heat, c_p, is expressed in cal/g · deg and the gas constant R = 1.987 cal/mol · deg, the adiabatic exponent, α, is a function of the atmospheric composition and humidity and can be determined using the following formulas:

$$\alpha = \frac{R}{\mu(c_p + C_w + C_r)}, \qquad \text{(Eq. 2.13)}$$

and

$$c_p = \frac{[\{p_{N_2} c_p(N_2)\} + \{p_{O_2} c_p(O_2)\} + \{p_{CO_2} c_p(CO_2)\} + \{p_{Ar} c_p(Ar)\}]}{p} \qquad \text{(Eq. 2.14)}$$

where the gas constant, R = 1.987 cal/mol · deg; μ is the molar weight of the atmospheric air mixture (for Earth $\mu \approx 29$); $p_{N_2} = 0.7551$; $p_{O_2} = 0.2315$; $p_{CO_2} = 0.00046$; and $p_{Ar} = 0.0128$ atm are the partial pressures of, respectively, nitrogen, oxygen, carbon dioxide, and argon (Voitkevich and Kokin, 1990); $p \approx 1$ atm is the summation of all the gas components in the atmosphere that were included in the calculation; $c_p(N_2) = 0.248$; $c_p(O_2) = 0.218$; $c_p(CO_2) = 0.197$; $c_p(Ar) = 0.124$ cal/g · deg are the heat absorbing capacities of nitrogen, oxygen, carbon dioxide, and argon at constant pressure (Khodakovsky, 1971). $C_w + C_r$ are the adjustment factors considering the total heating effect of moisture condensation C_w (in a humid atmosphere) and of absorbing the radiation heat from the Sun to the Earth C_r. Thus:

$$C_w + C_r = \frac{R}{\mu\alpha} - c_p \qquad \text{(Eq. 2.15)}$$

The best match of the theoretical temperature distribution in the Earth's troposphere with averaged empirical data occurs at $\alpha = 0.1905$. For the dry Earth's atmosphere air mixture, $c_p = 0.2394$ cal/g-deg. Thus, for a humid, IR-absorbing air of a real but averaged troposphere with the temperature gradient of 6.5 deg/km may be found by use of Eq. 2.15, $C_w + C_r = 0.1203$ cal/g · deg K. For planets of a different atmosphere, these examples should be understood as a description of any thermophysical or chemical processes resulting in the heat release or absorption ($C_w + C_r < 0$) within the troposphere.

To determine the characteristic temperature of a planet, one needs to determine the coefficients C_w and C_r:

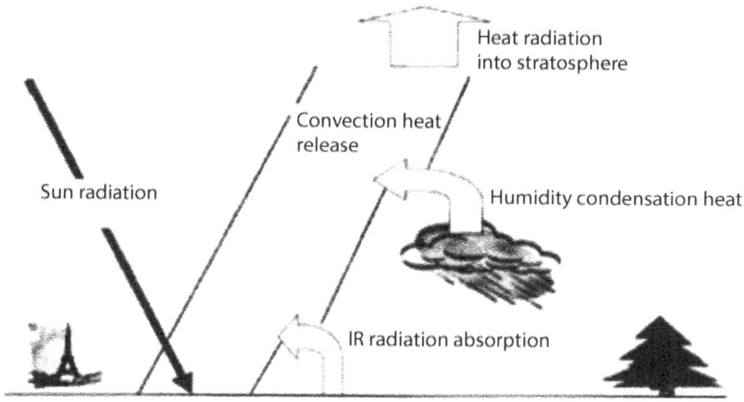

Figure 2.5 Balance of heat transfer in the Earth's troposphere. The loss with convective movements of air mass is 69%, moisture condensation adds another 25% and only 8% is the fraction of the radiation heat transfer. (After Sorokhtin *et al.*, 2011.)

$$C_r = \left\{\frac{R}{(\mu \cdot \alpha)}\right\} \cdot \left\{\frac{(T_s - T_e)}{T_s}\right\}, \qquad \text{(Eq. 2.16)}$$

and

$$C_w = \left\{\frac{R}{(\mu \cdot \alpha)}\right\} \cdot \left\{\frac{T_e}{T_s}\right\} - c_p \qquad \text{(Eq. 2.17)}$$

Inserting into Eqs. 2.16 and 2.17, the Earth's atmosphere parameter values are: $\alpha = 0.1905$, $\mu = 29$, $c_p = 0.2394$ cal/g · deg, $T_s = 288.2$ K, $T_e = 255$ K, and R = 1.987 cal/mole · deg one obtains $C_r = 0.0412$ cal/g · deg and $C_w + C_r = 0.1203$ cal/g · deg. Thus, one arrives at the same values as when determining these parameters from Eq. 2.15. This shows that a direct heat transfer by Earth's surface to the air mass participating in the tropospheric convective mass exchange reaches approximately 67%. The radiation component adds 8% to convective heat transfer and the heat release due to moisture condensation within the troposphere adds 25% more (see Figure 2.5).

Earth's Troposphere Model

A simplified equation for the tropospheric temperature, *T*, can be determined if one assumes: (1) $T_s = 288.2$ K; (2) the scaling factor as determined

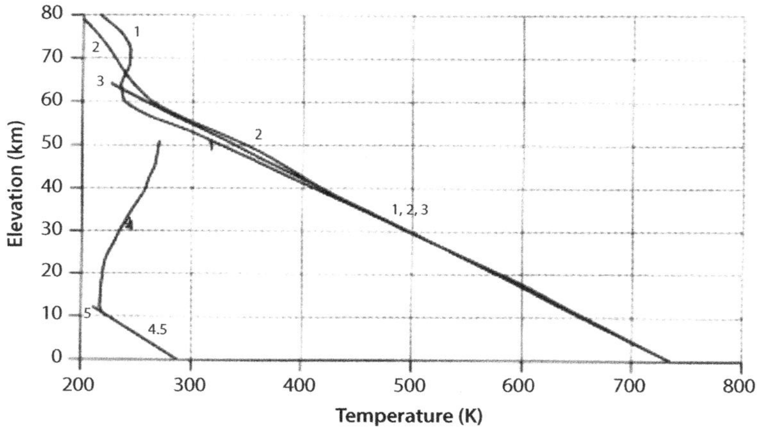

Figure 2.6 Temperature distribution in Earth's troposphere and stratosphere (*Curve 4*) and in the Venus troposphere (*Curve 2* at 30° latitude and *Curve 1* at 75° latitude) compared to theoretical temperature distributions (*Curves* 5 and 3) plotted in compliance with the greenhouse effect adiabatic theory, Eq. 2.25.

by Eq. 2.12 for the following conditions: ($\psi = 23.44°$, $T_e = 263.6$ K, $T_s = 288.2$ K, and at $\psi = 0°$; T_e=255 K, T_s; = 278.6 K); and (3) for the Earth a range of precession angles, $ba = 1.093$:

$$T_t = 288.2\left(\frac{p_t}{p_o}\right)^{\alpha} \qquad \text{(Eq. 2.18)}$$

At $\psi = 23.44$ °C, the best fit of the theoretical temperature, T_t, distribution in the troposphere and the averaged empirical data occurs at $a = 0.1905$ (see Figure 2.6).

Effect of Precession Angle

By definition, the *greenhouse effect*, ΔT, is the difference between the planet's average surface temperature, T_s, and its effective temperature T_e:

$$\Delta T = T_s - T_e \qquad \text{(Eq. 2.19)}$$

Average temperature for the entire Earth's surface is approximately 288 K or 15 °C. Its effective temperature at $\psi = 0$ is $T_e = 255$ K or −18 °C. Thus, the present-day value of the *greenhouse effect* for the Earth is 33 °C.

If we consider that the present-day Earth's precession angle is $\psi = 23.44°$, then the effective temperature of the Earth turns out to be $T_e = 263.49$ K (Sorokhtin, 2006).

The adiabatic model allows one to estimate the effect of the so-called *greenhouse gases* on the temperature regimes of the Earth's troposphere and the *greenhouse effect*. For asymptotic estimates, the writers assumed that the nitrogen-oxygen Earth's atmosphere was completely replaced first by carbon dioxide and, then, by methane, using the same atmospheric pressure, $p_s = 1$ atm or1.01324×10^2 KPa. The adiabatic exponent was determined from Eq. 2.14 at $\mu_{CO_2} = 44$, and $c_p = 0.197$ cal/g · °C, $\mu_{CH_4} = 16$, and $c_v = 0.528$ cal/gm · °C, thus, $\alpha_{CO_2} = 0.1423$ and $\alpha_{CH_4} = 0.1915$. Substituting these adiabatic, *a*, values into Eq. 2.13 with the same *b* factor value of 1.597, one can construct the temperature distribution for the hypothetical carbon dioxide and methane atmospheres. The corresponding near-surface temperature of the hypothetical carbon dioxide atmosphere would be 281.5 K, (6.4 °C) lower than that at the nitrogen-oxygen composition of the atmosphere, and for the methane atmosphere it will be 288.1 K, which is just 0.1 °C above the usual average Earth's temperature of 288 K. Thus, as the carbon dioxide atmosphere is denser and one, conversely lighter, the same pressures in these atmospheres will be attained at different elevations than those in the nitrogen-oxygen atmosphere:

$$h_{CO_2} = h_{N_2+O_2} \frac{\mu_{N_2+O_2}}{\mu_{CO_2}} \quad \text{(Eq. 2.20)}$$

and

$$h_{CH_4} = h_{N_2+O_2} \frac{\mu_{N_2+O_2}}{\mu_{CH_4}} \quad \text{(Eq. 2.21)}$$

where $\mu_{N_2+O_2} (= 28.9)$ is the molar weight of the nitrogen-oxygen atmosphere, and $\mu_{CO_2} (= 44)$ and $\mu_{CH_4} (=16)$ are the molecular weights of carbon dioxide and methane, respectively. The constructed temperature distributions within the hypothetical totally carbon dioxide and totally methane atmospheres are shown in Figure 2.7 together with the already presented the existing nitrogen-oxygen temperature distribution in the Earth's atmosphere (Figure 2.6).

Thus, for the hypothetical carbon-dioxide atmosphere with the same near-surface pressure of 1 atm, the average Earth's surface temperature

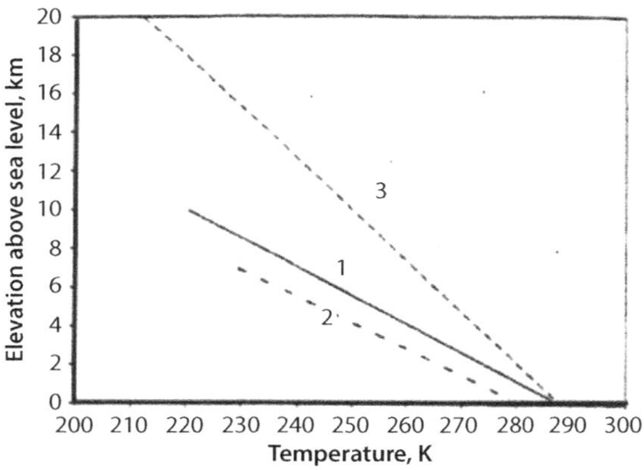

Figure 2.7 Averaged temperature distributions in the Earth's troposphere using Eq. 2.25: *Curve 1* – Earth's troposphere with a nitrogen-oxygen atmosphere; *Curve 2* – Earth's atmosphere composed totally of carbon dioxide; and *Curve 3* – Earth's atmosphere composed totally of methane. (After Chilingar *et al.*, 2009, p. 1210, figure 2.)

declines by approximately 6.5 °C (and not increasing significantly as commonly believed). Besides, due to a higher molecular weight of carbon dioxide, temperature within the entire thickness of such a troposphere is always lower than in a nitrogen-oxygen troposphere.

For a hypothetical methane atmosphere, the near-surface temperature at sea level remains almost unchanged, because $\alpha_{N_2+O_2} = 0.1905 \approx \alpha_{CH_4} = 0.195$. At the same time, in the troposphere it is higher than that for the nitrogen-oxygen atmosphere, because $\mu_{CH4}(=16) < \mu_{N_2-O_2}(=28.9)$, the methane atmosphere is much thicker than the nitrogen-oxygen one. That is why in the mountainous areas, surface temperature may significantly increase under such an atmosphere.

Similarly, on substituting a hypothetical nitrogen-oxygen for a carbon-dioxide Venetian atmosphere, at the same pressure of 90.9 atm, the Venetian surface temperature will rise from 735 to 795 K (462 to 522 °C; see Figure 2.7).

These estimates show that saturation of the atmosphere with carbon dioxide, with all other conditions being equal, results not in an increase, but rather a decrease of the *greenhouse effect* and the average temperature within the entire layer of the planet's troposphere. This happens despite intense absorption of the heat of radiation by CO_2. The physical explanation of this phenomenon is clear: molecular weight of carbon dioxide is 1.5 times higher and its heat-absorbing capacity is 1.2 times lower than those of the Earth's air. As a result (see Eqs. 2.14 and 2.18), the adiabatic exponent, for a carbon dioxide atmosphere, at the same conditions, is about 1.34 times lower than that for a nitrogen-oxygen humid air: $\alpha_{N_2+O_2} = 0.1905$.

From the thermodynamic viewpoint, the explanation of this phenomenon lies in the fact that the heat release from the troposphere occurs mostly due to air mass convection, which is a much more efficient mechanism, where the gas molecules are closer together, than the heat transfer by radiation. After the greenhouse gases absorb the heat of radiation, the energy of this radiation is converted into energy of thermal oscillations of gas molecules. This, in turn, leads to the expansion of the gas mixture and its rapid rise toward the stratosphere, where due to the rarified nature of the stratosphere, the excess heat is radiated into space.

Therefore, in a troposphere with an elevated carbon dioxide content the convection of the atmospheric gases will substantially accelerate.

Under the classic version of the *greenhouse effect* theory, the solar constant in Eqs. 2.3 and 2.5 are divided by four as the Sun-illuminated disk is four times smaller than the total surface of the revolving Earth. Strictly speaking, such action is valid only if the planet's revolution axis is perpendicular to the plane of the ecliptics, that is, if the precession angle $\psi = 0$. The precession angle, y, is the angle between the planet's revolution axis and the perpendicular to the ecliptics plane.

Examining an extreme situation of a planet *lying on its side*, its revolution axis is in the plane of the ecliptics and directed toward the Sun. The total area of the illuminated Earth is then somewhat lower than the visible Earth's circle, whereas when its revolution axis is perpendicular to the direction toward the Sun, the total planet's area is again four times the area of its illuminated circle. The determination of the effective temperature for the solar constant in Eq. 2.5 must be divided not by four, but by a value N where $4 > N > 2$. For intermediate cases, with the Earth's revolution axis inclined to the ecliptics, it is necessary to consider the proportionality of these two extreme cases to the inclination angle and its complement (for today's Earth, $N \approx 3.5$).

The Earth's equator is presently inclined to its orbital ecliptic at the angle $\psi = 23.44°$. Therefore, the areas beyond the polar circle are illuminated by the Sun on the average half a year and are devoid of the solar light during another half. While one near-polar area is illuminated by the Sun, another one is in the dark. Because of this, while calculating the average annual effective temperature for the polar areas, one should divide the solar constant by four, not by two. Taking sphericity of the Sun-illuminated polar area into account, the divisor should be $2(S_s/S_e)$, where S_s is the Earth's near-polar area beyond the polar circle and S_0 is the areal extent of the near-polar area projection onto the planet's equatorial plane:

$$\frac{S_s}{S_o} = 2\left[\left(\frac{1-\cos\psi}{(\sin\psi)^2}\right)\right] = \left[\frac{2}{(1+\cos\psi)}\right] \quad \text{(Eq. 2.22)}$$

For the other areas of the Earth's regularly illuminated by the Sun, the Eqs. 2.3, 2.5 and 2.10 are valid. The Earth's effective temperature, T_e, can be determined using the following equation (Sorokhtin, 2006):

$$T_e = \left[\frac{S(1-A)}{\sigma\left\{4\left(\frac{\frac{\pi}{2}-\psi}{\frac{\pi}{2}}\right)+4\left(\frac{\psi}{\frac{\pi}{2}}\right)\left(\frac{1}{1+\cos\psi}\right)\right\}}\right]^{\frac{1}{4}} \quad \text{(Eq. 2.23)}$$

Because the current precession angle of the Earth is $\psi = 23.44^\circ$, the factor in the denominator of Eq. 2.23 is:

$$4\left[\left\{\frac{\left(\frac{\pi}{2}-\psi\right)}{\frac{\pi}{2}}\right\}+\left\{\frac{\psi}{\frac{\pi}{2}}\right\}\left\{\frac{1}{1+\cos\psi}\right\}\right] = 3.502 \quad \text{(Eq. 2.24)}$$

For an Earth's albedo, $A = 0.3$, the effective temperature as calculated by Eq. 2.23 is $T_e = 263.6$ K. The tropospheric temperature, including the surface temperature, may be expressed through the planet's effective temperature:

$$T_t = b^\alpha\left[\frac{S(1-A)}{\sigma\left\{4\left(\frac{\frac{\pi}{2}-\psi}{\frac{\pi}{2}}\right)+4\left(\frac{\psi}{\frac{\pi}{2}}\right)\left(\frac{1}{1+\cos\psi}\right)\right\}}\right]^{\frac{1}{4}}\left(\frac{p_t}{p_o}\right)^\alpha \quad (2.25)$$

Empirically, the average near-surface Earth temperature at a pressure of $p = p_0 = 1$ atm and $\psi = 23.44^\circ$ is approximately 288 K or +15 °C (The Atmosphere, 1988; Reference Book, 1951).

Using today's values for the Earth, e.g., the present-day precession angle at $\Psi = 23.44^{o}$, $A \approx 0.3$, $\alpha = 0.1905$ and the near-surface pressure at the sea level $p_s = 1$ atm, the proportionality coefficient, *b*, as defined by Eq. 2.10, can be determined based on a condition that the present day average Earth's surface temperature is 288 K or +15 °C (Bachinsky *et al.*, 1951). Then $b = 1.597$ and for the oxygen--nitrogen atmosphere $b^{\alpha} = 1.093$. At a constant albedo, *A*, but a different composition of the atmosphere, *b*, (coefficient value for Earth) remains the same, whereas b^{α} will change depending on the value of the adiabatic exponent α describing the atmosphere composition.

Table 2.1 was developed using Eq. 2.25, the equation used for estimating Earth's troposphere temperature distribution, T_t, based on a pressure set taken from the standard atmosphere model (Bachinsky *et al.*, 1951). These results were based on the present-day precession angle $\Psi = 23.44^{o}$, assuming $p_s = p_0 = 1$ atm as shown in Figure 2.6. Table 2.1 shows that the estimates for the Earth agree with the temperature distribution in the tropospheric model of the standard Earth atmosphere within an accuracy of about 0.1%. The standard Earth atmospheric model in its substance is averaged over the entire Earth based upon correlation of the temperature and pressure vs. the elevation above the sea level. This tropospheric model with the gradient of 6.6 K/km is customarily used for tuning aviation altimeters and calibrating barometers intended for surface observations.

Application of Adiabatic Equation to the Planet Venus

A more rigid check of the Eq. 2.25 is its application to another planet, Venus. The estimation of temperature distribution for Venus's troposphere of a different atmosphere of dense carbon gases is based on actual measurements of its atmospheric pressure and composition. The temperature for the Venus troposphere was measured by using satellites. Two sets of data are available for this study: (1) the one presented in Planet Venus (1989): $p_s = 90.9$ atm and $T_s = 735.3$ K; and (2) data presented by Marov (1986): $p_s = 95$ atm, $T_s = 740$ K, $S = 2.62 \times 10^6$ erg/cm$^2 \cdot$ s, $A = 0.77$ and $\Psi \approx 3^{o}$. It was assumed that $A = 0.76$ and $\Psi \approx 3.18^{o}$. Using a similar technique for Venus as was done for the Earth, the best fit of the theoretical temperature distribution with its empirical values occurs at the adiabatic exponent $a = 0.1786$ and ba factor of 1.429 ($b = 7.37$). For Venus, $c_p = 0.2015$ cal/g · °C, $T_s = 735.3$ K and $T_e = 230.5$ K. Then, $C_r = 0.1756$ cal/gm · °C, $C_w = 0.1213$ cal/g · °C and $C_q = (C_r + C_w) = 0.0543$ cal/g · °C. The C_r

Table 2.1 Temperature distribution in the Earth's troposphere and tropopause based\upon the Earth's standard atmosphere.

Standard atmospheric model (Bachinsky, 1951)					**Theoretical estimate (Eq. 2.25)**		
***h*, km**	***p*, mm Hg**	***T*, °C**	***T*, K**		***p*, atm**	***T*, K**	***T*, °C**
0.0	700.00	15.00	288.00		1.00	288.00	15.00
0.5	716.01	11.75	284.75		0.9421	284.75	11.77
1.0	67411	8.50	281.50		0.8870	281.50	8.50
1.5	634.21	5.25	278.25		0.8345	278.24	5.24
2.0	596.26	2.00	275.00		0.7846	275.00	2.00
2.5	560.16	−1.25	271.75		0.7371	271.74	−126
3.0	525.87	−4.50	268.50		0.6919	268.49	−431
3.5	493.30	−7.75	265.25		0.6491	265.24	−7.76
4.0	462.40	−11.00	262.00		0.6084	262.00	−11.00
4.5	433.10	−14.25	258.75		0.3699	285.75	−14.25
5.0	405.33	−17.50	255.50		0.3333	255.49	−17.51
5.5	379.04	−20.75	252.25	Troposphere	0.4987	252.25	−20.75
6.0	354.16	−2400	249.00		0.4660	249.01	−23.99
6.5	330.72	−27.25	245.75		0.4345	245.71	−27.29
7.0	308.52	−30.50	242.50		0.4059	242.55	−30.45
7.5	287.55	−33.75	239.25		0.3784	239.33	−33.67
8.0	267.79	−37.00	236.00		0.3524	236.10	−36.90
8.5	240.16	−40.25	232.75		0.3278	232.87	−40.13
9.0	231.62	−43.50	229.50		0.3048	229.67	−13.33
9.5	215.09	−46.75	226.25		0.2830	226.44	−46.56
10.0	199.60	−50.00	223.00		0.2626	223.24	−19.75
10.5	185.01	−53.25	219.75		0.2434	220.03	−52.97
11.0	171.34	−56.50	216.50		0.2254	216.84	−56.16
11.5	160.11	−56.50	216.50	Tropo-pause	0.2107	214.07	−58.93
12.0	I40M	−56.50	216.50		0.1969	211.32	−61.68

Table 2.2 Venus troposphere: empirical temperature and the temperature estimated from Eq. 2.25.

Empirical data from Planet Venus, 1989					**Theoretical calculated data, Eq. 2.25**		
***h*, km**	***T*, K**	***p*, bar**	***T*, K**	***p*, bar**	***h*,km**	***p*, atm**	***T*, K**
0	735.3	92.10			0	90.92	736.9
1	727.7	86.45			1	85.34	728.6
2	720.2	81.09			2	80.05	720.3
3	712.4	76.01			3	75.03	712.1
4	704.6	71.20			4	70.29	703.8
5	696.8	66.65			5	65.79	695.5
6	688 8	62.35			6	61.55	687.3
7	681.1	58.28			7	57.53	679.1
8	673.6	54.44			8	53.74	670.9
9	665 8	50.81			9	50.16	662.7
10	658.2	47.39			10	46.78	654.4
12	643.2	41.12			12	40.59	638 1
14	628 1	35.57			14	35.11	621 8
16	613.3	30.66			16	30.27	605.5
18	<97.1	26.33			18	25.99	589.2
20	580.7	22.52			20	22.23	573
22	564.3	19.17				18.92	556.8
24	547.5	16.25			24	16.04	540.6
26	530.7	13.70			26	13.32	524.3
28	513.8	11.49			28	11.34	508.1
30	496.9	9.581			30	9.458	491.9
Measurement latitude 0–30"			Meas. latitude. 75'		Theoretical estimate		
33	471.7	7.211	471.7	7.211	33	7.118	467.6
36	448.0	5.346	4465	5.345	36	5.277	443.2
39	425.1	3.903	420 5	3.894	39	3.848	418.9
42	403.5	2.802	394.5	2.78	42	2.755	394.6
45	385.4	1.979	368.7	1.941	45	1.935	370.5
48	366.4	1.375	3433	1.321	48	1.331	346.6
51	342.0	0.9347	318.5	0.8741	51	0.8905	322.6
54	312.8	0.6160	290.0	0.3582	54	0.5796	298.7
57	282.5	0.3891	258.2	0.3392	57	0.3595	274.3
60	262.8	0.2357	2373	0.1918	60	0.2125	249.7

parameter determines the absorption of the planet's heat radiation by atmosphere. Its relatively elevated temperature value is due to a high atmospheric pressure and very hot troposphere. One can assume $C_w < 0$ for the Venus troposphere where endothermic reactions predominate, especially in its lower and middle layers. There is dissociation of some chemical compounds, for instance, dissociation of the sulfuric acid into $SO_3^=$ and water: $H_2SO_4 + Q \rightarrow SO_3^= + H_2O$, where $Q \approx 15$ kcal/mol. For the upper layer of the Venetian troposphere at the elevations between 40 and 60 km, $C_w > 0$. Therefore, the exothermic chemical reactions, e.g., the reduction of sulfuric acid, and condensation of water vapor in the clouds prevail. The best match of the theoretical curve with empirical data occurred at $a = 0.1786$, coefficient $b = 7.372$ and $b^a = 1.429$.

Again, a check was conducted using Eq. 2.25. The data from the check is included in Table 2.2 and is shown in Figure 2.6. As shown in Table 2.2, the theoretical temperature distribution in the Venus atmosphere shows a good match with the experimental measurements as quoted in Planet Venus (1989). Up to an elevation of 40 km, the accuracy is in fractions of a percentage point, whereas at the elevations between 40 and 60 km, theoretical temperatures are positioned between two series of empirical data. At higher elevations, $p < 0.2$ atm, the tropopause begins in the Venus atmosphere, and the adiabatic theory no longer applies as convection is no longer the primary transfer of heat.

It should not be considered accidental that a match of the derived temperature distributions is found for the troposphere's of the Earth and Venus. This is a solid confirmation of the validity of Eq. 2.25. This equation may be used for determination of the Earth's climates in the past geologic epochs and their forecast for the future. However, for this purpose it is necessary to review the evolution of the Earth's atmospheric composition and pressure of the atmospheric gases in the following chapters.

3

The Earth's Synoptic Activities

The term synoptic activity (scale) refers to large-scale processes that have a horizontal length scale on the order of 1000 kilometers (about 620 miles) or more. This corresponds to a horizontal scale typically found in mid-latitude atmospheric depressions, e.g., extratropical cyclones. Most high and low meteorological areas seen on weather maps, such as surface weather analyses, are synoptic-scale systems, driven by air pressure in their respective hemisphere. The word *synoptic* is derived from the Greek word συνοπτικός, *synoptikos*, meaning *seen together*.

Greenhouse Effect Adiabatic Theory

The *greenhouse effect* adiabatic theory referred to in Chapter 2 can be defined as a unidimensional model. The planet would be represented as a dimensionless point and the only dimension is the elevation, h, above this point (sea level). Such a model is an accurate determination of the global tropospheric parameters, such as its: (1) *greenhouse effect* (average surface temperature), (2) average value of the radiation or (3) humidity-condensational component of the tropospheric heat release.

Lambert's law can be used to describe the illumination of a sphere and converted into a multi-dimensional model. Addition of the longitudinal component and the seasonal fluctuations of the planet's illumination converts this model into a three-dimensional model. A four-dimensional model is created if one adds time. These additions, however, could decrease the model's accuracy in expressing the *greenhouse effect* correlation with the composition of the planetary atmosphere.

Around 1729, Lambert stated that the absorbance of a material is directly proportional to the thickness (path length) it passes through. In 1852, Beer recognized another attribute of this formula, that the absorbance is proportional to the concentrations of the attenuating species. Today, the Beer–Lambert law combines the two laws and correlates absorbance to both the concentrations of the attenuating species as well as the thickness of the material sample. This law is often referred to as Lambert's Law.

Lambert's Law can be stated as, the absorbance of light is directly proportional to the thickness of the media through which that light is being transmitted, multiplied by the concentration of absorbing chromophore:

$$A = ebc \qquad \text{(Eq. 3.1)}$$

where A is the absorbance, e is the molar extinction coefficient, b is the thickness of the solution, and c is the concentration. This law tends to break down at higher concentrations, because at higher concentrations, the molecules are closer to each other and begin to interact with each other. This interaction between the molecules alters several properties of the molecules, thus changing the attenuation. In the case of intense radiation, nonlinear optical processes can cause variance. Lambert's law can be further modified to include the latitude, ψ, which controls the concentration of the incoming light.

Model of Heat Transfer in the Troposphere

It is fundamental to recognize that most heat within the troposphere is exchanged by convection and not radiation (see Chapter 2). Lambert's law enables us to model the transfer of heat in the atmosphere. Included in the proposed model is the latitude, φ. The physical definition of the temperature of an *absolutely-black-body* must be replaced by the concept of temperature for a *gray-body*, T_{gb}:

$$T_{gb}^4 = \frac{S}{4\sigma}\cos\varphi \qquad \text{(Eq. 3.2)}$$

where φ is the location's latitude.

Considering the convective heat transfer in the Earth's troposphere, the Earth-imitating *gray-body* temperature may be expressed as follows:

$$T_{gb}^4 = \left(\frac{S}{4\sigma}\cos\,\psi + \frac{\dot{Q}}{\sigma}\right)^{\frac{1}{4}} \qquad \text{(Eq. 3.3)}$$

where $\dot{Q}$ is the rate of heat transfer by the air masses (erg/cm^2 · s), e.g., cyclones. The tropospheric temperature, accounting for consideration of the Sun and the Earth's precession, under the accepted approximation as determined in Chapter 2 will be:

$$T = b^{\alpha}\left[\frac{S(1-A)}{\sigma\left(4\dfrac{\left(\dfrac{\pi}{2}-\psi\right)}{\dfrac{\pi}{2}} + 4\left(\dfrac{\psi}{\dfrac{\pi}{2}}\right)\left(\dfrac{1-\cos\psi}{(\sin\psi)^2}\right)\right)}\cos\varphi_{\psi} + \frac{\dot{Q}}{\sigma}\right]^{\frac{1}{4}}\left(\frac{p}{p_o}\right)^{\alpha}$$

(Eq. 2.25)

where $\varphi_{\psi} = (1-\psi\delta)$ is the effective latitude of the location, in consideration of the fact that the Earth's polar areas are illuminated by the Sun only half a year and are in the dark the other half; δ is the delta function ($\delta = 1$ in summer for a given hemisphere, whereas $\delta = 0$ in the winter for the same hemisphere). Determination of the average annual values is $j\psi$ $\delta = ½$.

Eq. 2.26 enables the determination of a latitudinal zoning of the near-surface temperatures, T_s. If, however, a latitudinal distribution of empirically measured average annual temperature is available, then it is possible to determine the average specific rate of heat input for an air mass, $\dot{Q}$, at a given latitude!

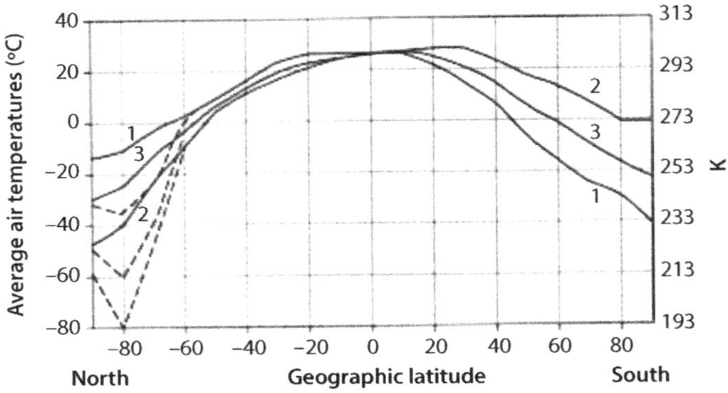

Figure 3.1 Near-surface average air temperature versus latitude. Solid lines, after Khromova and Petrosyants (2001). *Curve 1* – January; *Curve 2* – July; and *Curve 3* – annual average temperatures. Dashed lines (Antarctic data) are average temperatures (obtained from "The Atlas of Antarctic", 1966), for the ice cover dome at the Antarctic pole (80 °S). This data was from well observations in 1964 of the annual average temperature (–60 °C) by Sorokhtin and A. Kapitsa. (After Sorokhtin *et al.*, 2011, figure 13.10, p. 490.)

Eq. 2.26 enables the calculation of the synoptic processes evaluation rate of $\dot{Q}$ if the near-surface temperature, T_s, latitudinal distributions are available. If however, a latitudinal distribution of empirically measured annual temperature is available, then it is possible to determine average specific speed of the heat input with the air mass $\dot{Q}$ at a given latitude. Figure 3.1 displays the average near-surface air temperature vs. geographic latitude (Khromov and Pertosyants, 2001). This figure shows the surface temperature, T_s, theoretical distribution for various $\dot{Q}$ values. These $\dot{Q}$ values are selected on condition of the best T_s theoretical distribution match with empirical data. Figure 3.2 includes the diagrams of the Earth's troposphere synoptic activity vs. latitude. This correlation was determined by variations of the average specific rate of heat input for an air mass, $\dot{Q}$, parameter in Eq. 2.26 on condition of the best match between the Earth near-surface temperature values (as determined from the same equation at $p = p_0$) and its empirical values provided in Figure 3.1.

Figures 3.1 and 3.2 show that the energy release-rate by the trade winds is slightly approximant to 4 to 4.5 × 10^4 erg/cm^2 · s. In the Northern Hemisphere, the average energy release-rate by the air mass cyclonic activity gradually increases toward the high latitudes. It appears that the air mass motion energy is transmitted from the tropical belt to the northern boreal and polar areas. For instance, the local maximum in the rate of synoptic

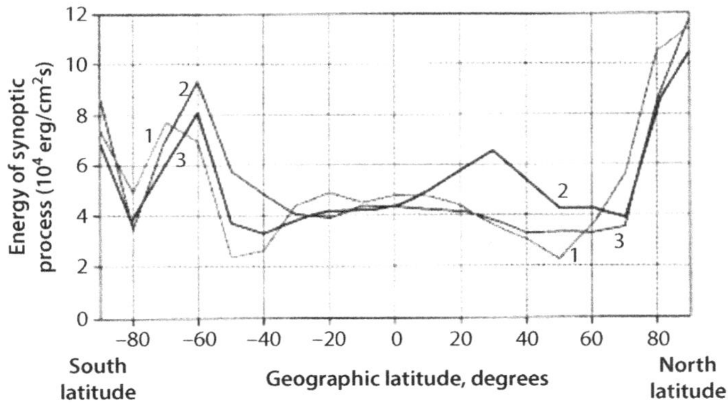

Figure 3.2 Synoptic process activity vs. latitude calculated using Eq. 2-25: *Curve 1* – January; *Curve 2* – July; *Curve 3* – annual average. (After Sorokhtin *et al.*, 2011, Figure 11.10, p. 497.)

energy release in July at 30° N latitude (see Figure 3.2) indicates the emergence of the summer tropical storms and hurricanes. They all occur at that time, for instance, in the northern areas of the Caribbean Basin, and in the Pacific: in the Philippines, in the South Japanese Islands and southern China. It is possible that New Orleans' destruction by Katrina in 2005 was exactly due to such hurricanes. Reflected in the diagrams, significant synoptic activity growth in the Northern Hemisphere polar areas is in good correlations with conclusions of many climatologists. As an example, Khromov and Pertosyants (2001) indicated that in the Arctic basin, in all seasons, intense cyclonic activity is observed. Cyclones emerge on Arctic fronts and slide in the Arctic from lower latitudes where they develop on polar fronts.

Substantial increase in the synoptic activity in the polar areas of the Northern Hemisphere is shown on these diagrams, which also agrees with the views of the leading climatologists. For instance, Khromov and Petrosyants (2001) indicated that intense cyclonic activity is observed in the Arctic basins in all seasons. The cyclones emerge on the Arctic fronts and penetrate the Arctic from the lower latitudes where they develop on the polar fronts.

The Antarctic synoptic activity was measured at the South Pole, Antarctica, approximately along the meridian 90°E along a profile from the Mirny Station through the continent's highest elevation (the ice sheet dome at 80°S and at an elevation of 4000 m above sea level). The Antarctic is almost continuously dominated by anti-cyclonic conditions with clear and quiet weather. Usually cyclones can penetrate only the nearshore areas; however, synoptic activity distribution in central Antarctica turned out very peculiar. Minimum activity was observed over the dome and a drastic increase in the energy of the synoptic process activity was observed southward, toward the pole, and northward. The question is, "Why do observed bursts of synoptic activity emerge over the continents?"

An explanation of this paradox is that there is a very strong katabatic wind continuously blowing from the Antarctic continental high dome toward the shores. In Antarctica, these winds are especially strong in winter (in July), as observed by Dr. O. Sorokhtin. These strong (up to 45 mph) cold winds become warmer adiabatically as they descend from the continental dome on the ice-covered slopes. The air mass descending from the continental dome carries a substantial amount of energy. The maximum release-rate occurs at the continental shores around a circle about the 60° South latitude. Outside of the continent, the katabatic winds disperse over the expanses of the Southern Ocean and gradually die out. The katabatic winds that cause a *bicephalous* (double-headed) shape of the synoptic activity over the Antarctic.

Figure 3.3 shows the latitudinal correlations of average semi-annual Sun radiation energy flows and the synoptic processes. Average value of the solar insolation was 2×10^5 erg/cm$^2 \cdot$ s, whereas the average synoptic process energy is equal to 0.39×10^5 erg/cm$^2 \cdot$ s. The total amount of solar energy reaching the Earth is about 10^{24} erg/s, whereas the total

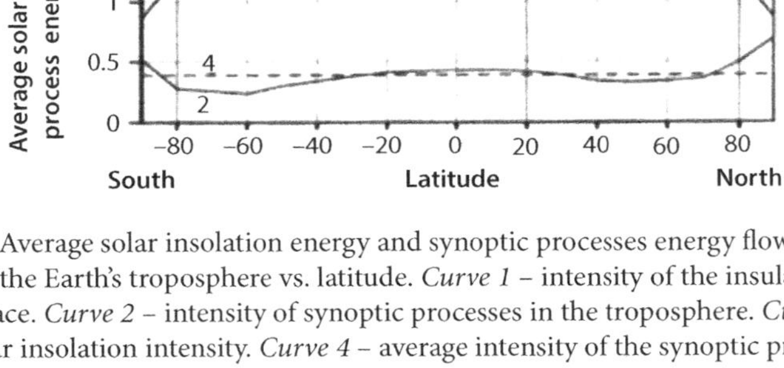

Figure 3.3 Average solar insolation energy and synoptic processes energy flow (105 erg/cm^2/sec) in the Earth's troposphere vs. latitude. *Curve 1* – intensity of the insulation at the Earth's surface. *Curve 2* – intensity of synoptic processes in the troposphere. *Curve 3* – average solar insolation intensity. *Curve 4* – average intensity of the synoptic processes.

energy of synoptic processes is only 0.2×10^{24} erg/s. Obviously, against this synoptic process energy background, the total energy generation by mankind is negligibly low, 1.3×10^{20} erg/s. *Consequentially, this is another reason why the anthropogenic energy effect on the Earth's climate should be disregarded.*

For an Earth with a greater snow covering, the albedo (A_s) increases. Under these conditions, the Earth's near-surface temperature may be estimated using Eq. 3.4 for the planet's average albedo $A \approx 0.3$, replaced by a higher value of $A_s \approx 0.7$. This explains such phenomena as, for instance, the chilling of near-surface air layers in clear nights under anticyclones, when $S \approx 0$ and the heat input $\dot{Q}$ is low. Indeed, for anticyclonic areas the rate of convective air mass exchange usually slows down. As a result, the convective heat input declines; however, the heat radiation from the Earth's surface heats up during the day. As a result, the $\dot{Q}$ factor in Eq. 2.26 declines and at night time, at $S \approx 0$, the near-surface temperature substantially declines. In winter, in high latitudes where the Earth's surface is covered by the snow with a higher albedo, heating by the solar radiation is insignificant, and results in air overcooling and bitter cold. For stable anticyclones, $\dot{Q} \approx 0$, in snow-covered regions a general troposphere overcooling occurs with the atmospheric tropopause descending almost to the Earth's surface. Clear examples of such air overcooling are conditions arising in the Central Antarctic and in winter in the Yakutia and Verkhoyansk areas of Russia. When the anticyclonic regime in the troposphere is replaced by cyclonic activity, there is an immediate resumption of the input of synoptic energy and convective air mass mixing. Warming occurs, and on the average reviewed adiabatic temperature distribution approximately is restored.

Generalizing the model as a 4-D version, it is necessary to introduce the longitude angle and identify the oceanic and continental areas. A 4-D model requires the introduction of seasonal and daily variations in the Earth's illumination, etc. Thus, the proposed model allows a derivation of the local climatic features of the planet. Earth's surface albedo, heat introduction by cyclones, and atmospheric humidity must be included in this model. In areas of high-snow-cover reflectance but devoid of the heat input from cyclones, the Earth's surface temperature declines almost to the tropopause (lower stratospheric stratum) temperature, which is defined by the atmosphere's radiation balance for a given latitude (Figure 3.4). In summer, in such anticyclonic areas with a dry air, the other way around, overheating of the near-surface tropospheric strata it occurs approximately by 4 to 5 °C and higher, with all indications of a drought, e.g., Russia's Trans-Volga steppes (see Figure 3.4). The temperature gradient calculation for a dry

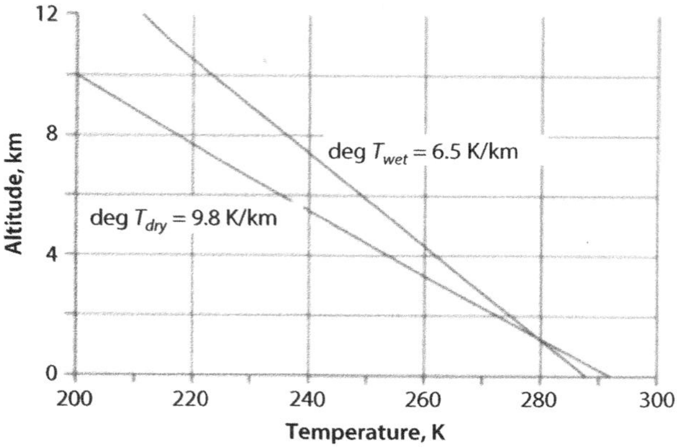

Figure 3.4 Comparison of temperature distributions plotted using Eq. 2.25 for a dry, transparent (grad T_{dry}) and humid, IR radiation--absorbing the Earth's troposphere (T_{wet}). For all other conditions being equal, the near-surface temperature of a humid, absorbing troposphere is always somewhat lower than the near-surface temperature of a dry transparent atmosphere (in this particular example the temperature difference reaches 4.7 °C).

(anticyclonic) and humid (cyclonic) troposphere is usually performed using a simple equation:

$$\text{Grandient } T = \frac{g}{c_p} \qquad \text{(Eq. 3.5)}$$

where g is the gravitational acceleration and c_p is the specific heat.

Part II

DEVELOPMENT OF THE HYDROSPHERE

4

Development of Earth's Hydrosphere

The Earth's hydrosphere and atmosphere evolutionary regimes were crucial in the: (1) determination of the evolution of the Earth's climate; (2) the emergence and development of life on the Earth; and (3) the issue of commercial mineral deposits including the organic hydrocarbon deposits.

Hydrosphere of the Primordial Earth

When the Earth first formed it did not have either a hydrosphere (ocean, rivers, lakes, etc.) or atmosphere (see Figure 1.1). It is suggested that the external, and very mobile geosphere, primarily developed due to the degassing of the Earth's mantle. This degassing did not begin until after the heating of the upper portion of the Earth and the emergence of melted matter nodes within it. There was heating because of the release of energy from the moon-Earth interaction tidal energy and radioactive decay. Tidal energy was released primarily within the Earth's mantle. The first melts of the Earth's mantle material appeared at relatively shallow depths of 200–400 km. Upon the appearance of the first melts, the differentiation of the Earth's matter began, along with the first indications of the tectono-magmatic activity of the primitive Earth, ≈4 BY ago. After that, the matter differentiation of the Earth was fed by a powerful gravitational energy process of high-density melted iron separation from the Earth's matter silicates. The Earth's mantle-degassing substantially depended not only on the mantle temperature, which determined the mantle convection flow intensity, but also on its chemical composition. The key features of this compositional and petrogenic evolution of a convecting mantle chemical composition are illustrated in Figures 4.1 and 4.2.

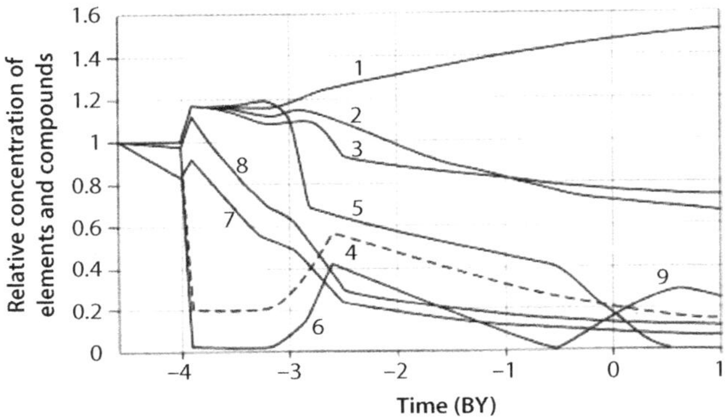

Figure 4.1 Evolution of convecting mantle's chemical composition in relative concentrations (the concentration of a given compound in the primordial Earth's matter is taken as one: *Curve 1* – SiO_2, TiO_2, MgO, CaO, AlO_3; *Curve 2* – H_2O; *Curve 3* – K_2O; *Curve 4* – Ni and other siderophilic and chalcophilic elements and compounds; *Curve 5* – FeO; *Curve 6* – Fe; *Curve 7* – U; *Curve 8* – Th; and *Curve 9* – Fe_3O_4. (After Sorokhtin *et al.*, 2011, figure 4.15, p. 144.)

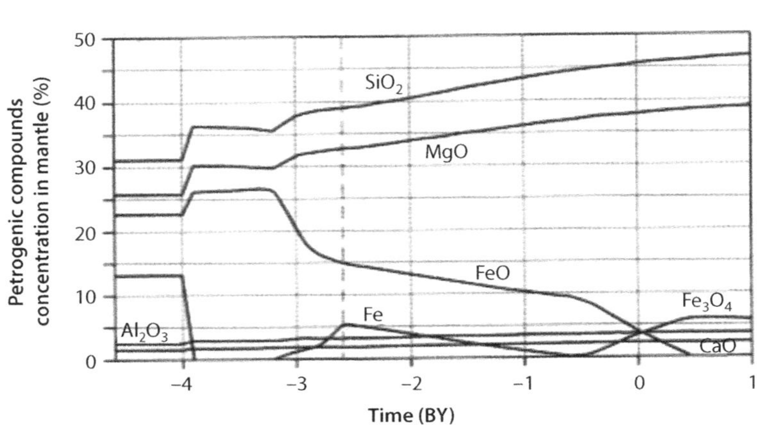

Figure 4.2 Evolution of major petrogenic elements and compounds in the convecting mantle. (After Sorokhtin *et al.*, 2011, figure 4.16, p. 145.)

The emergence of the Earth's lithospheric plate tectonics, especially the development of the Earth's evolution, provides an opportunity for a quantitative description of the ocean-forming processes. Quantitative models of the hydrosphere (global ocean water) and the atmosphere (gas shell) growth are based on the most general concept for the global evolution (Sorokhtin, 1974). This concept of development includes the direct proportionality between the Earth's degassing rates and the lithospheric plate tectonics as to the development of the lithosphere and hydrosphere. The major contribution to the mantle convective mass exchange came from the most powerful energy process, the chemico-density differentiation of the Earth's matter into a high-density iron-oxide core and a residual silicate mantle.

Earlier studies were related to the end of the Earth's formation process ≈4.6 BY ago; however, later models by Monin and Sorokhtin (1984), Sorokhtin and Ushakov (1991 and 2002) were more sophisticated and considered that the primordial Earth after its emergence from the collation of particles was a relatively cold planet. Thus, the degassing would have started, after the preliminary heating of the originally cold Earth, to a temperature where the silicates began to melt in the upper mantle along with the emergence of the first asthenosphere.

The primary mantle degassing is associated with the solubility decline of the volatile components within the silicate melts under lowering temperature and relatively low pressure. As a result, the mantle melts erupted on the surface of the Earth (mostly basalts, and in the Archaean time also komatiite lavas) boiled and released excess volatile elements and compounds into the atmosphere. Some of these volatile components may have been released due to the weathering of the erupted rocks after their sojourn on the surface. However, the main water-degassing mechanism for the Earth was the decline of its solubility under the cooling and crystallization of water-containing olivine basalt melts at low pressure (Figure 4.3).

The rate of the Earth's degassing was in direct proportion with: (1) the erupted mantle rock mass on the surface, per unit of time; (2) their content of volatile elements; and (3) their mobility. For a first approximation, the mantle rock eruption rate may be considered to be proportionate to the Earth's tectonic activity (Figure 4.4). This activity is determined by the rate of its total heat loss, $\dot{Q}_m$, or the derivative over time of the Earth's tectonic parameter, $\dot{z}$:

$$\dot{z}_n = \frac{(\dot{Q}_m - Q_{4.0})}{(Q_{mo} - Q_{4.0})}, \qquad \text{(Eq. 4.1)}$$

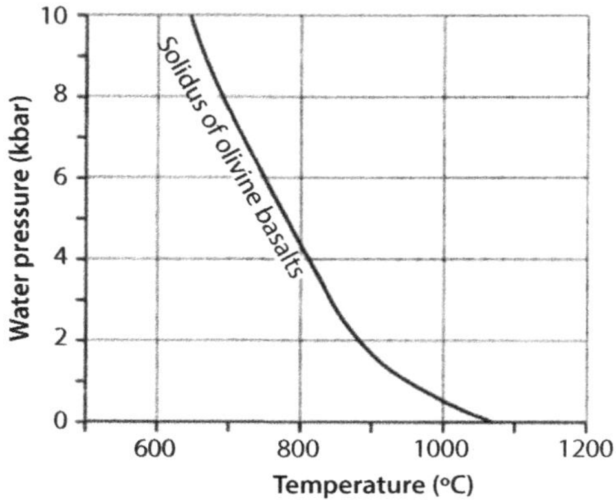

Figure 4.3 Olivine basalt solidus curve versus pressure (content) of water dissolved in the basalt melt. When basalts crystallize, the water dissolved in the basalt melts is released. (After Joly 1929; in: Sorokhtin *et al.*, 2011, p. 422, figure 11.1.)

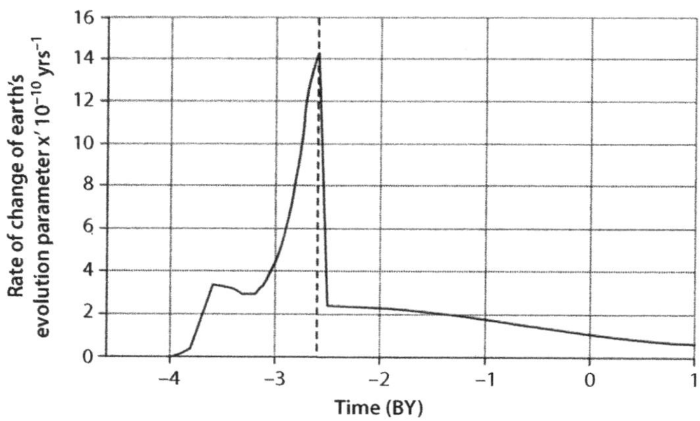

Figure 4.4 Rate of change in the Earth's tectonic parameter $\dot{z}$ determining the rate of the basalt melts surface eruptions and the mantle degassing rate (the dashed line represents core separation time). (After Sorokhtin *et al.*, 2011, Figure 11.2, p. 423.)

where $\dot{Q}_m$ is the current value of the heat flow from the mantle, $Q_{4.0} \approx 1.6 \times 10^{37}$ erg is the Earth's heat loss at the start of its tectonic activity 4.0 BY ago and $Q_{mo} \approx 10.77 \times 10^{37}$ erg is today's total heat loss of the Earth's mantle. Then the derivative $\dot{z}$ (Eq. 4.1) and its normalized value

$$\dot{z}_n = \frac{\dot{Q}_m}{|\dot{Q}_{mo}|},$$

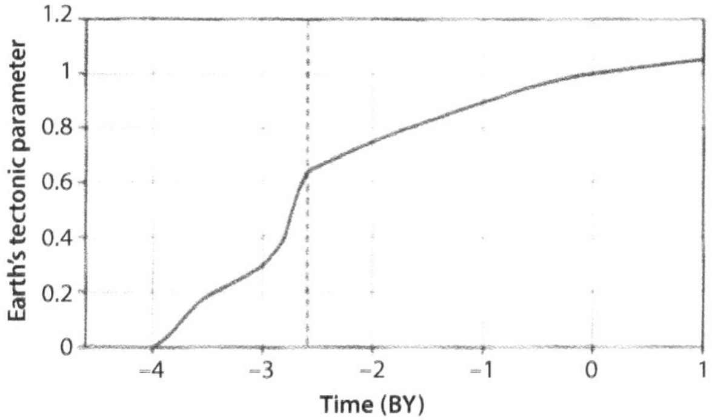

Figure 4.5 Earth tectonic parameter, $\dot{z}$, controlling the mantle magmatic activity products accumulation rate and the mass of the substances degassed from the mantle: the dashed line represents the time core separation ends. (After Sorokhtin *et al.*, 2011, Figure 11.3, p. 423.)

where $\left|\dot{Q}_{mo}\right| = 3.39 \times 10^{20}$ erg/s is the absolute value of today's Earth deep heat flow. The correlation of the Earth's tectonic parameter change rate versus time is shown in Figure 4.4 and the value of the parameter controlling the magmatic activity is shown in Figure 4.5.

Thus, the mantle degassing rate is in direct proportion with the component content in the mantle m_{i_m}, its mobility factor, χ_i, and the rate of the mantle convective mass exchange $\dot{z}$:

$$\dot{m}_i = -m_{i_m} \chi_i \dot{z} \tag{Eq. 4.2}$$

The mass of an ith degassed volatile component in the mantle and its accumulation in the Earth's external geospheres is determined as:

$$m_i = m_{i_o}(1 - e^{-\chi_i z}), \tag{Eq. 4.3}$$

where m_{i_o} is the total mass of the ith volatile component in the mantle and external geospheres. During the Archaean time, the degassing of the mantle mass gradually increased. Thus, it is necessary to consider its increase from $M_m = 0$ to its total amount at the end of Archaean, M_{mS}. The mantle mass was previously estimated in Figure 4.6. It is necessary to use a different equation for the degassing of the water or any other volatile element and compound (such as N_2 and CO_2) in the Archaean time:

$$m_i = m_{i_o}(1 - e^{-\chi_i z})\left(\frac{M_m(t)}{M_{m\Sigma}}\right) \tag{Eq. 4.4}$$

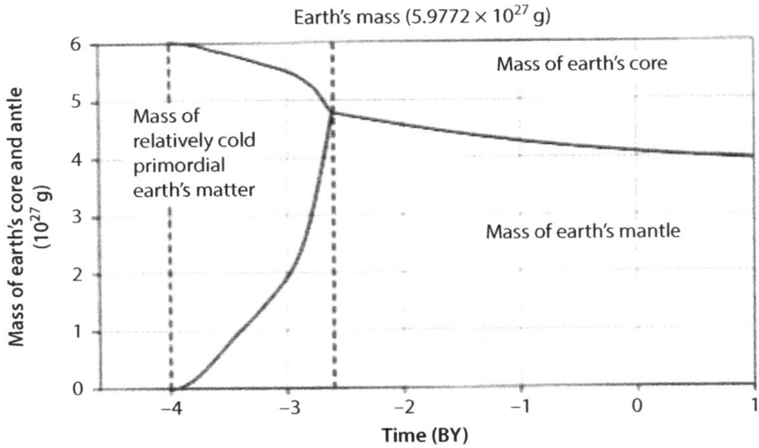

Figure 4.6 Gradual increase of the mantle mass in the Archaean time due to the expansion of the Earth's matter differentiation zone and in its mass in Proterozoic and Phanerozoic due to the core growth. Dashed lines are the times when the core matter separation began about 4 BY ago and when a high-density iron-oxide core separated in the center of Earth about 2.6 BY ago. Since the origin of the Earth 4.6 BY ago and through the very end of Archaean time, the primordial Earth's matter saturated with ore elements was preserved within Earth. (After Sorokhtin *et al.*, 2011, figure 6.6, p. 208.)

where $M_m(t)$ is the mantle mass at the time $4 < t < 2.6$ BY ago. For the Proterozoic and Phanerozoic time, Eq. 4.3 is valid.

To determine the mass of the volatile component m_i, degassed from the mantle (e.g., water), it is necessary to insert in Eqs. 4.3 and 4.4 the initial and boundary conditions for the content of this component (water, nitrogen, and/or carbon dioxide) in the Earth's external geospheres.

Formation of the Hydrosphere

The Earth's matter differentiation mechanisms were drastically different from today's in the Archaean and post-Archaean time. Thus, it is to be expected that the mobility factors χ_i of the volatile components (or at least some of them) were significantly different.

During the Archaean time, the silicate matter of the convecting mantle, together with its volatile components, went through a layer of melted iron (see Figure 4.7). The oxides, with their lower heat of formation, rather than the oxide of bivalent iron (63.64 kcal/mol) must have dissociated and released their oxygen for the oxidation of iron to the bivalent oxide. The water (both vapor or fluid) molecule heat of formation is 57.8 kcal/mol.

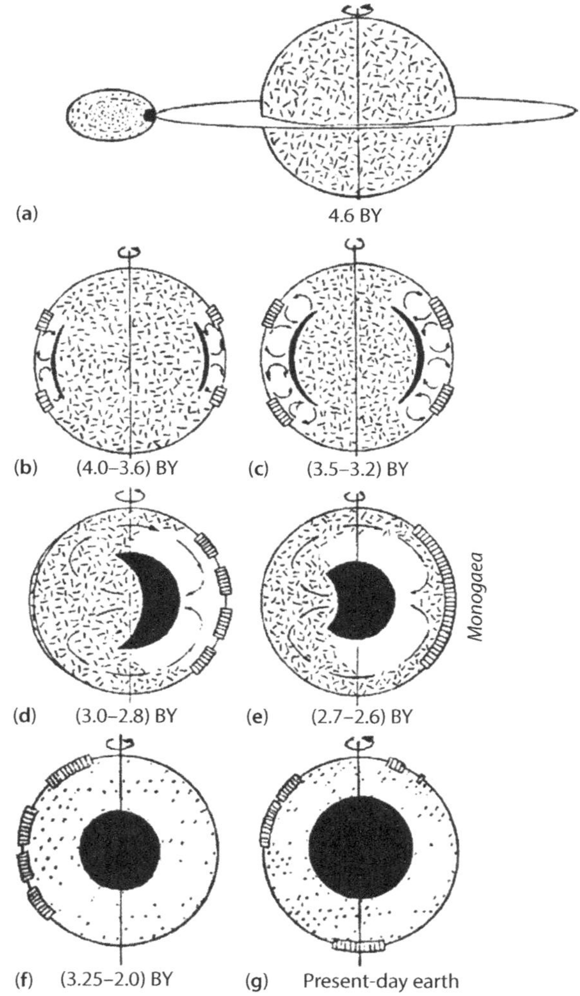

Figure 4.7 Earth structure evolution: (A) the young Earth and moon formation; (B–F) consecutive stages of the Earth's core separation and formation; (G) present-day Earth. The dashes represent the primordial matter; solid black is iron and its oxides melts; white is the Archaean depleted mantle impoverished in iron, its oxides, and siderophilic elements; the dots are the current-type mantle; and the boxes are continental massifs. (After Sorokhtin *et al.*, 2011, figure 4.1, p. 117.)

Therefore, the water must have dissociated on the metallic iron melts in the Earth's matter differentiation zones:

$$Fe + H_2O \rightarrow FeO + H_2 + 5.84\ \text{kcal/mol}. \qquad \text{(Eq. 4.5)}$$

The carbon dioxide heat of formation is 94.05 kcal/mol. Thus, carbon dioxide could have crossed through this zonal differentiation layer without difficulty. In the Archaean time, the water mobility factor in the degassing Eqs. 4.2 and 4.3 was significantly lower than that in the post-Archaean time, whereas the carbon dioxide mobility factor could remain constant during the entire time of its degassing from the mantle. It is noteworthy that together with the water, many other oxides with low formation heat dissociated in Archaean time on the iron melts. They were reduced to free elements.

The water mobility factors in Archaean time and thereafter significantly differed. Thus, to estimate the mantle degassing it is necessary to put together two degassing equations, Eq. 4.3 and 4.4, with different mobility factors. Then combine them by the degassing process continuity condition at the Archaean-Proterozoic time boundary. In this case, these two equations include three variables: two water mobility factors and the original water mass ($m_{H_2O_0}$) in the Earth's matter.

To solve the problem quantitatively, it is necessary to determine and insert into the equations three independent boundary conditions.

1. ***The first boundary condition*** can be the total mass of water in the present-day ocean and the continental crust. Our estimates are based on publications by Ronov and Yaroshevsky (1978) and amended by Sorokhtin *et al.* (2011), whose observations suggest the following assumptions for which the mass of water in the hydrosphere may be assumed: (1) The ocean currently contains 1.42×10^{24} g of water (in the continental crust together with the continental waters and icebergs of 0.446×10^{24} g. (2) The oceanic crust contains approximately 0.358×10^{24} g of irreducible water. (3) Altogether, the Earth's external geospheres (the hydrosphere) contain 2.23×10^{24} g of water. This amount of water was degassed from the Earth's depths over its geological life, that is, over the last 4.5 BY. Strictly speaking, this is not exactly accurate because a significant part of the water that reached the surface in Proterozoic and Phanerozoic time was constantly returned into the mantle through the plate subduction zones. Some portion of it dissociated in hydration of the oceanic crust rocks and in the upper atmosphere (affected by the solar radiation). The authors assume that not the absolute mass, but only its effective value is equal to the difference between the masses of the degassed and the

subducted water. In this situation, all calculations are valid and only the effective value of the mobility factors X will be somewhat lower than their actual values.

2. ***A second boundary condition*** is the total water mass on Earth, $(m_{H_2O})_o$. The first step is to determine to calculate its mass in the present-day mantle. The water concentration in the mantle is a fundamental problem considered by petrologists. It has not been conclusively resolved. The reason is that upon reaching the Earth's surface, practically all the mantle water immediately (during the rise and discharging onto the surface) is intensely contaminated by the surface water. Even the most depth-associated xenoliths, e.g., such as garnet peridotites or eclogites from the kimberlite blow pipes, turn out to be just fragments of the oceanic crust sucked-in to great depths underneath the continents through the former subduction zones (Sorokhtin, 1985; Sorokhtin *et al.*, 1996, 2004). Nevertheless, fresh basalts of the oceanic islands, despite a possibility that they capture some marine water, which filter through the stratovolcanoes, usually contain very little water, no greater than 0.3% OH (Joder and Tilly, 1962).

   This and other theoretical considerations lead most modern petrologists dealing with the origin of mantle rocks to believe that the mantle water content is exceptionally low: Ringwood (1981) assumed the mantle water content at 0.1% and Pugin and Khitarov (1978) believed in a lower value of 0.025 to 0.1%. Sorokhtin and Ushakov (1991, 2002) estimated values no greater than 0.05 to 0.06%. After a review of all estimates, one can conclude that the mantle is exceptionally dry. The total water content in today's mantle is $\approx 2.007 \times 10^{24}$ g and the total water mass on Earth, $(m_{H_2O})_o \approx 4.23 \times 10^{24}$ g.

3. ***A third boundary condition*** for the calculations could be the hydrosphere's water mass at an intermediate point in time (if it is possible). Using additional geological data, it is a solvable problem. As the ocean gradually increased in volume, there should have been a moment in its evolution when the oceanic water covered the mid-oceanic ridge crests with their associated rift zones. After that, hydration of the oceanic crust rocks must have been rapidly increasing and so must have been the release of the ore element from the rift zones to the ocean.

Therefore, the geochemistry of oceanic deposits must have drastically changed after the described event as they now include abundant ore elements brought out from the mantle.

The most characteristic such element and the clear indicator of the sought-for boundary (the moment of the oceanic crust saturation with water) is iron. Metallic (free) iron was still available in the Precambrian mantle in noticeable amounts (see Figure 4.2). Rising together with the hot mantle matter into the rift zones, it reacted with the marine water (devoid of oxygen) rich in carbon dioxide forming bicarbonates of iron:

$$Fe + H_2O + 2CO_2 \rightarrow Fe(HCO_3)_2 \quad \text{(Eq. 4.6)}$$

After the ocean covered the mid-oceanic ridge crests, the dissolved $Fe(HCO_3)_2$ began spreading all over the ocean. In the shallow water, the bivalent iron helped by bacteria (especially cyanobacteria) was oxidized to the trivalent form and precipitated forming thick iron ore deposits:

$$2Fe(HCO_3)_2 + O \rightarrow Fe_2O_3 + 2H_2O + 4CO_2 \quad \text{(Eq. 4.7)}$$

Because of the iron release from the oceanic rift zones, at least two unique massive iron ore formation epochs are identified in the World Ocean history. The first was at the end Archaean time (the Kivatin-type iron ore formations) 3 BY ago. The second was at the end of the Early Proterozoic time, nearly 2.2 BY ago, known as the Krivoy Rog-type ores. This second iron accumulation stage is most identifiable in the Earth's geological record.

The area of the oceans in Proterozoic is the global area minus the continental area. The continental crust thickness has had minor change with respect to time; therefore, the area of the continents can be considered as proportional to their mass. The mass evolution of continents is presented in Figure 4.8. The area of the oceans 2.2 BY ago was $\approx 3.48 \times 10^{18}$ cm^2. The average ocean depths at that time was ≈930 m. Thus, the World Ocean water mass 2.2 BY ago was $\approx 0.325 \times 10^{24}$ g (Figure 4.9). Knowing the bound water mass in the crust 2.2 BY ago, it is possible to calculate, using a reconstruction (Sorokhtin and Ushakov, 2002) of the development of the oceanic crust structure (Figure 4.10) and the continental crust mass at that time (see Figures 4.8). It may be estimated that by the end of Archaean time, the average irreducible water concentration in the continental crust rose to 0.6%. Today it is 2%. Thus, 2.2 BY ago the continental crust contained $\approx 0.109 \times 10^{24}$ g of water. Assuming the average density of the pelagic deposits at ≈2.2 g/cm^3 with 20% water, the hydrated basalts and gabbros (density ≈2.9 g/cm^3) containing ≈2.5%, and the serpentinites (density

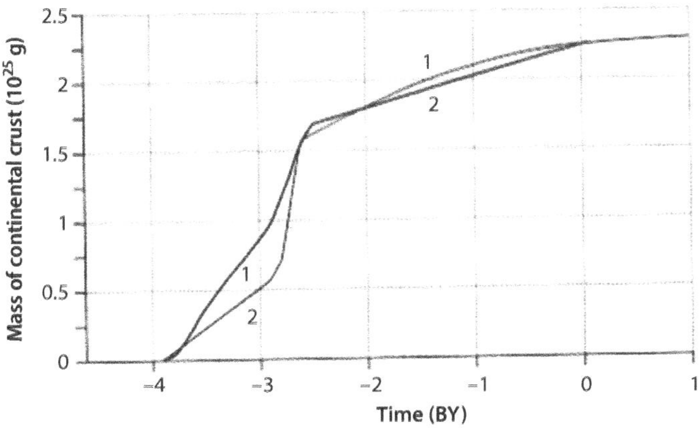

Figure 4.8 Continental crust mass growth: (1) authors' version and (2) Taylor and McLennan (1985) curve. (After Sorokhtin *et al.*, 2011, figure 7.7, p. 250.)

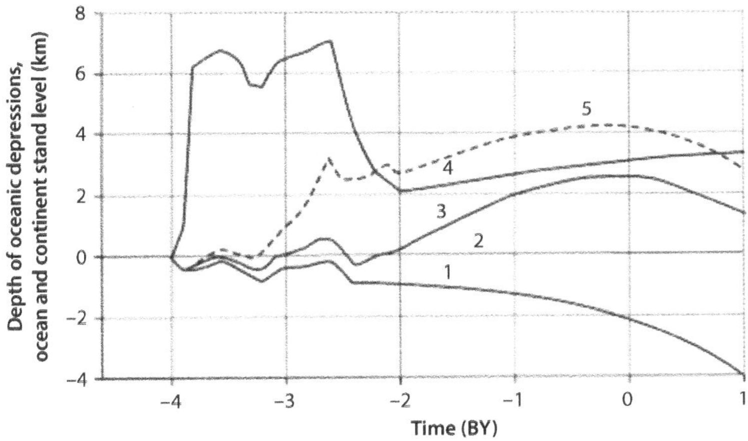

Figure 4.9 Evolution in the positions of the oceanic and continental surfaces compared with the average mid-oceanic ridge crests stand level: *Curve 1* – average depth of the oceanic depressions; *Curve 2* – mid-oceanic ridge crests stand level; *Curve 3* – world-ocean surface level; *Curve 4* – average continent stand level (relative to mid-oceanic ridge crests stand level); and *Curve 5* – ocean surface position in the case of no water dissociation on iron melts in the Archaean time under the reaction: $Fe + H_2O = FeO + H_2 + 5.84$ kcal/mol. (After Sorokhtin *et al.*, 2011, figure 11.6, p. 428.)

$\approx$3 g/cm^3) up to 11%, 2.2 BY ago the oceanic crust contained $\approx 0.385 \times 10^{24}$ g of water. Thus, the total water mass in the hydrosphere was:

$$(0.325 \times 10^{24}) + (0.385 \times 10^{24}) + (0.109 \times 10^{24}) \approx 0.819 \times 10^{24}\ \text{g}. \qquad \text{(Eq. 4.8)}$$

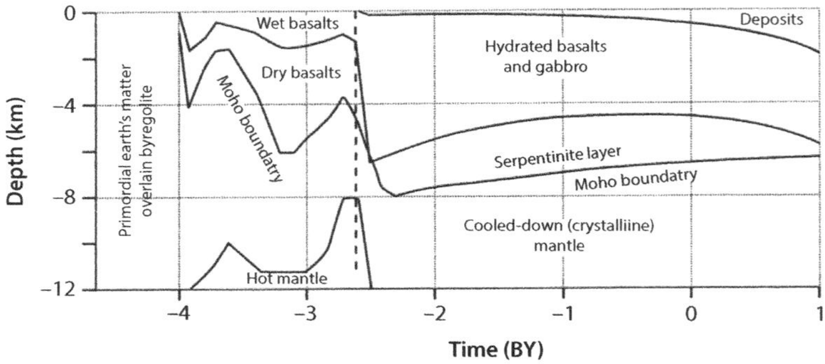

Figure 4.10 Generalized model of the oceanic crust evolution. (After Sorokhtin *et al.*, 2011, figure 7.1, p. 243.)

Constructing a system of Eqs. 4.3 and 4.4 separately for the Archaean and the post-Archaean time, one can substitute into them today's water masses in the Earth, in its hydrosphere, and in the Early Proterozoic hydrosphere ≈2.2 BY ago. These equations have different water mobility factors. Thus, they need to be merged into a common system through the condition of degassing continuity at the Archaean-Proterozoic time boundary. After that the correlation of the mantle degassed-water versus time can be determined. The water mobility factor in the Archaean time turns out to be $\chi_1 = 0.123$ and in the Proterozoic and Phanerozoic it was almost 12 times greater, $\chi_2 = 1.45$.

The water accumulation in the Earth's hydrosphere is illustrated in Figure 4.11. This figure demonstrates that the water accumulation regime in the external geospheres changed substantially near the Archaean-Proterozoic time boundary. The change was especially drastic for the water accumulation in the oceanic crust, which caused the formation of the serpentinite layer of the oceanic crust at the beginning of Proterozoic. This layer is the most capacious reservoir of the irreducible water on Earth.

After determining the total water mass in the Earth and the water mobility factors for the Archaean and post-Archaean time, using Eq. 4.3 one can determine the water accumulation rate in the Earth's hydrosphere. The water-from-mantle to hydrosphere degassing rate is shown in Figure 4.12. This graph indicates that the maximum water degassing rate occurred 2.5 BY ago, which is early in Proterozoic time. At the same time, maximum tectonic activity had occurred during the Archaean. This apparent discrepancy was caused by the fact that in the Archaean time, most of the

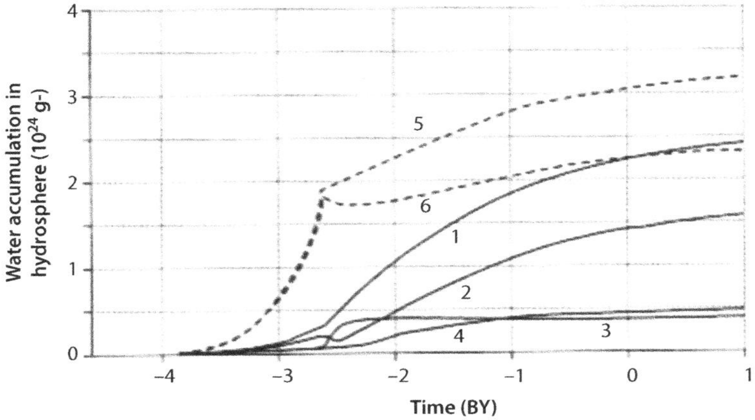

Figure 4.11 Water accumulation in the Earth's hydrosphere: *Curve 1* – total mass of water degassed from the mantle; *Curve 2* – oceanic water mass; *Curves 3 & 4* – water mass bonded in the oceanic and continental crust; and *Curves 5 & 6* – water mass degassed form the mantle and the oceanic water mass under a hypothetical scenario of no water dissociation in the Earth's matter differentiation zones in the Archaean time according to the reaction Eq. 4.5. (After Sorokhtin *et al.*, 2011, figure 11.4, p. 427.)

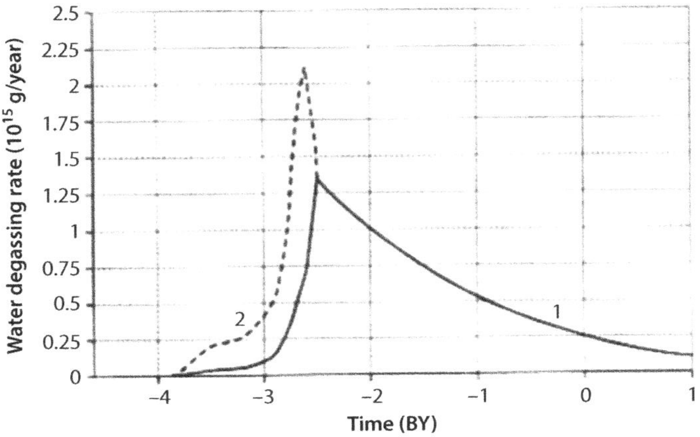

Figure 4.12 Mantle-to-hydrosphere water degassing rate: *Curve 1* – accounting for water dissociation in the Earth's matter differentiation zones in the Archaean time and *Curve 2* – not accounting for such dissociation. (After Sorokhtin *et al.*, 2011, figure 11.5, p. 427.)

degassed water was still within the mantle, dissociated over the iron melts in the Earth's matter differentiation zones as follows:

$$H_2O + Fe \rightarrow FeO + H_2. \qquad \text{(Eq. 4.9)}$$

The heavy hydrogen isotope is mostly associated with iron, whereas the light hydrogen isotope was found in the degassed water. In the Proterozoic time and even more so in Phanerozoic time, there was no water dissociation in the mantle. According to Eq. 4.2, water (without any losses) entered the Earth's hydrosphere. Currently the yearly rate of water degassed from the mantle is $\approx 2.6 \times 10^{14}$ g/year or 0.26 km^3/year.

The average depth of oceanic depressions may be estimated from the Earth's tectonic activity (see Figure 4.13) and from the average oceanic plate longevity (see Figure 4.14). Determination of the average depth of

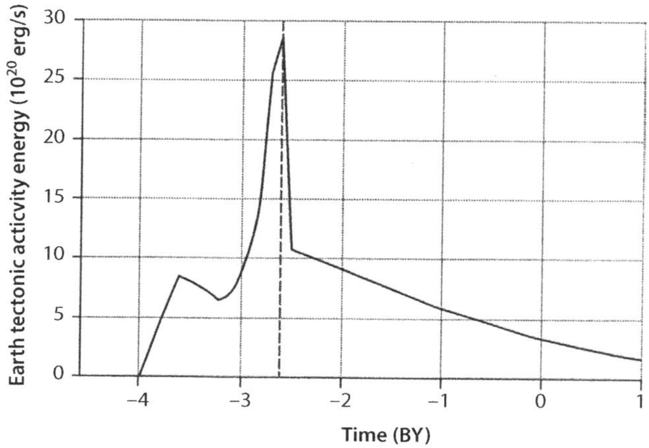

Figure 4.13 Energy expression of the Earth's tectonic activity (vertical dashed line denotes the core formation time). (After Sorokhtin *et al.*, 2011, figure 5.17, p. 187.)

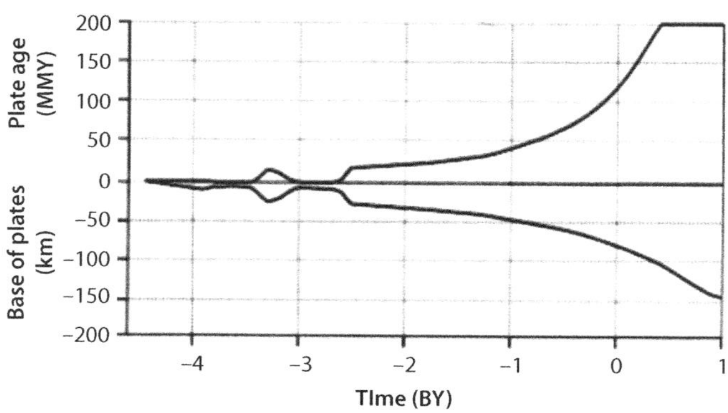

Figure 4.14 Evolution of oceanic lithospheric plates and their average thickness. (After Sorokhtin *et al.*, 2011, figure 7.3, p. 245.)

the oceanic depressions and their areas can be estimated, keeping in mind that the oceans in the Early Archaean time were positioned only in the low latitudes. It is now possible from the determined mass of water mass in the ocean to obtain the ocean surface elevation relative to the average standing level of the mid-oceanic ridge crests. The results are displayed in Figure 4.9.

Figure 4.9 establishes that the oceanic depression depths, relative to the average standing level of the mid-oceanic ridge crests, in the Early Archaean time were small: 80 to 200 m. There wasn't a large quantity of water in the oceans during this period. During the Early and Middle Archaean time, there were still no real oceans on the Earth. What existed were numerous isolated shallow-water marine-type basins. During this period, the level of these seas rose over the mid-oceanic ridge crests and the Earth's primordial matter areas, but especially the lithospheric plate pileup zones (Figure 4.15). The kernels of future continental massifs were being formed in the centers of these zones. Their elevations reached 6 km (see Figures 4.9 and 4.16).

The high continent stand level in the Archaean time was determined by the high heat flows. This resulted in thick and heavy lithospheric plates, which could not form underneath the heavy continental shields. The continental crust, because of a lighter density (buoyancy), rose high over the average mantle surface. Today, the present-day continental crust is underlain by a thick (up to 200 km) and dense (≈ 3.3 g/cm^3 ultramafic lithosphere), which substantially sinks the continents into the upper mantle (see Figure 4.17).

The high stand of the continental shield during the Archaean time resulted in intense physical erosion of its surface (Sorokhtin *et al.*, 2011). This is observed in many of today's Archaean shields on whose surface are deeply exposed, metamorphosed amphibolite and granulite facies that were formed at depths of 5–10 km. There are also examples of deeply metamorphosed epidote-amphibolite facies, e.g., in the Kola Peninsula Keiva

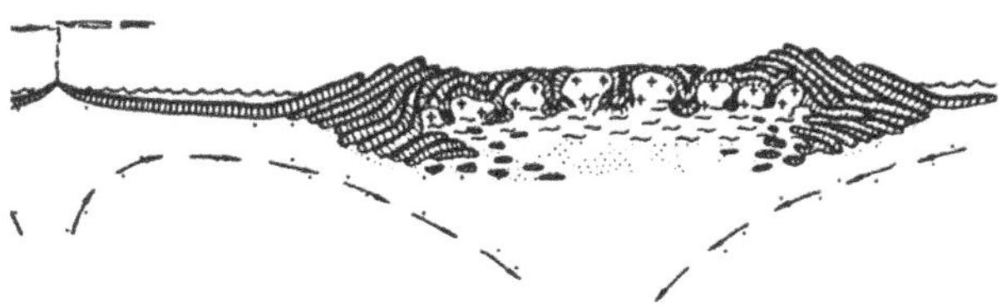

Figure 4.15 Image of the continental crust formation in Archaean time. (After Sorokhtin *et al.*, figure 7.5, p. 247.)

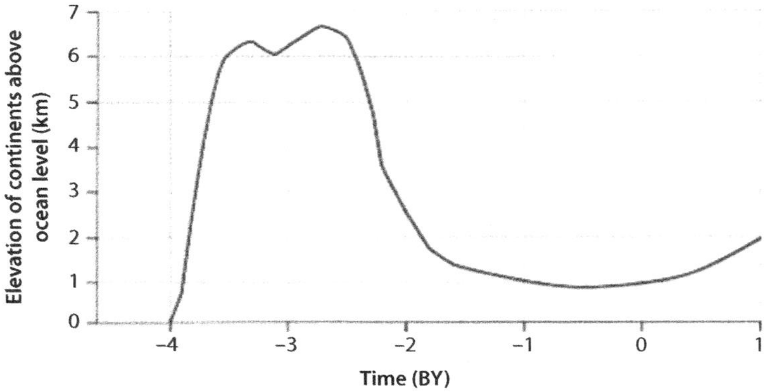

Figure 4.16 Average stand (elevation) of continental massifs over the rift zones in the Archaean time and over the ocean surface in Proterozoic and Mesozoic. (After Sorokhtin *et al.*, 2011, figure 7.10, p. 353.)

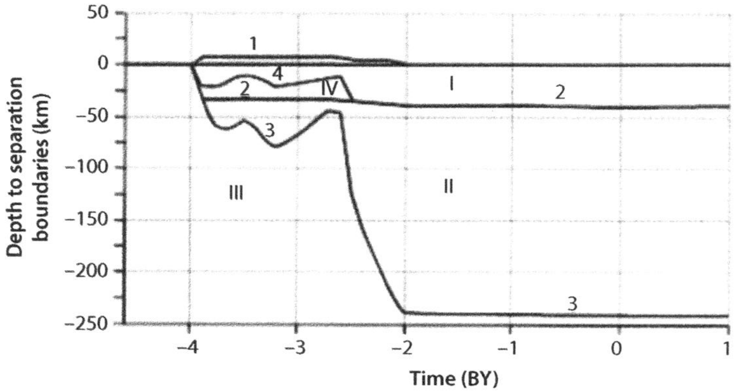

Figure 4.17 Evolution of the structure of the continental plates: *Area I* – continental crust; *Area II* – continental lithosphere; and *Area III* – sublithospheric (hot) mantle. *Curve 1* – surface of the continents; *Curve 2* – continental crust base (Mohorovicic boundary); and *Curve 3* – base of the continental lithosphere. (After Sorokhtin *et al.*, 2011, figure 7.9, p. 252.)

block. Huge masses of sedimentary rock had to be deposited on the ocean floor during the high stand of the Archaean continental blocks, after they remelted in the oceanic plate pileup zones and converted to massive granitoid intrusions of Archaean granite-greenstone belts Figure 4.15).

At the end Proterozoic, the surface of the growing ocean had risen to the average elevation of the continental planes. In the Phanerozoic, the first global marine transgressions on the continents occurred, which noticeably shrank the river discharge and continental rock erosion dump areas in the oceans.

If water dissociation occurred during the Archaeon time within the Earth's matter differentiation zone, then the mass of water in the hydrosphere would have exceeded the current amount by 1.5 times: $\approx 3 \times 10^{24}$ g rather than the $\approx 2.2 \times 10^{24}$ g (see Figure 4.11). The ocean surface elevation would then be about 2 km above its current elevation. This would have resulted in the flooding of the Earth in the Archaean and Protozoic time over a large portion of today's continental areas (see Figure 4.9). Only the highest continental areas would have been above the water surface.

Part III

DEVELOPMENT OF THE EARTH'S ATMOSPHERE

5

Earth's Historic Atmospheres

Earth's Primordial Atmosphere

Our solar system originated from the debris of explosions of earlier Suns and was formed from some of that debris. The Sun and its associated planets that were formed from this solar nebula, coalesced ≈ 4.6 BY ago. The initiation of hydrogen fusion into helium gave birth to our Sun. The planets orbiting the Sun were also formed by accretion; the heavier elements from the nebula were concentrated within their cores and the lighter elements, e.g., H_2 and He became the initial atmosphere. Figure 5.1 reflects that we lost much of our initial atmosphere, e.g., H_2, He, CH_4 etc. during the Hadean time. The geologic history of the composition of the Earth's various atmospheres with respect to time is presented in Table 5.1 and Figure 5.1.

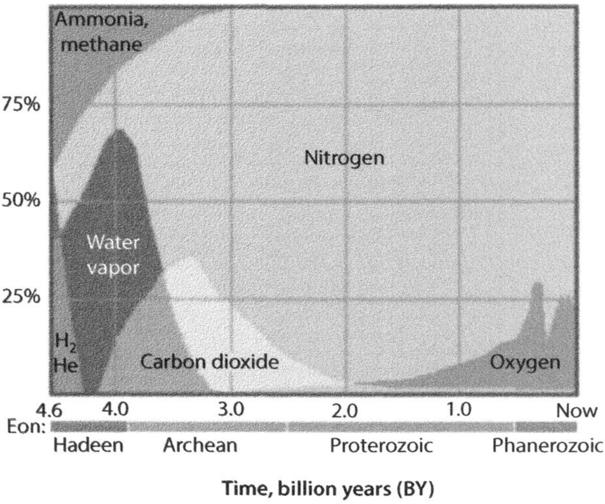

Figure 5.1 Variation of the composition of the Earth's atmosphere with time. (After Scientific Psychic, 2017.)

Table 5.1 Evolution of the Earth's Atmospheres.

	Atmospheric composition	Source of its components	Other characteristics
1st Atmosphere Hadean time	CH_4, He and H_2	Asteroids, comets, etc.	Atmosphere mainly H_2 and He. Volatile compounds tripped out
2nd Atmosphere Archean and 1st half of Proterozoic time	CO_2, H_2O and methane (decreasing) and N_2	Mantle degassing	Primarily CO_2 and N_2 by end of Archean, presence of bacteria producing N_2.
3rd Atmosphere 1st half of Proterozoic and Phanerozoic time	Primarily N_2 and O_2 with small amounts of Ar, CO_2, etc.	O_2 generated by photosynthesis	Development of ozone layers

Earth's First Atmosphere (Hadean time—4.56 to 4.0 BY ago)

The initial atmosphere formed when the material forming the Earth accreted and melted, forming layers of material with the heavier compounds toward the core of the Earth. The surface of the Earth was covered by a layer of ultrabasic porous regolith. This layer consisted of a loose, heterogeneous superficial material covering the solid rock of the Earth's surface, which included dust, soil, broken rock, and other related materials. This regolith material absorbed all chemically active gases, e.g., CO_2, CO, O_2, or H_2O. As a result, the Earth's initial primordial atmosphere primarily consisted of (1) H_2, (2) He, (3) nitrogen (N_2), and (4) a trace of noble gases (Figure 5.1). The atmospheric pressure was ≈1 atm.

At the beginning of the Hadean time (4.6 BY ago) the Earth's surface consisted of molten rock and water existed as vapor in the atmosphere (see Figure 5.1). Because of the elevated temperature, the average speed of the hydrogen molecules (H_2) and the helium atoms (He) was greater than their escape velocity from the Earth. In the Early Hadean time, due to Earth's weaker gravity, these initial lighter gases escaped from the Earth's atmosphere and were swept away from the Earth through photoevaporation by the solar winds. At the end of the Hadean time, the Earth's atmosphere was left with less methane and ammonia, water vapor, and small percentages of nitrogen and carbon dioxide as shown in Figure 5.1. Scientific Psychic (2017) has noted that at the end of the Hadean time, there was likely a cataclysmic meteorite bombardment (≈3.9 BY ago) that kept much of the Earth's surface in a molten state. The incoming impactors likely brought additional water, methane, ammonia, hydrogen sulfide and other gases that supplemented the Earth's atmosphere. High surface temperatures of 700 °C to 1100 °C during the Hadean time were common on the hot Earth's crust along with lakes of magma on its surface. These elevated temperatures were the result of: (1) collisions and compression during accretion, (2) decay of short-lived radioactive elements and (3) constant volcanism. Water only existed in gaseous form in the atmosphere and the Earth's atmosphere was depleted of methane through the following endothermic reaction:

$$CH_4 + H_2O \rightarrow CO + 3H_2 \qquad \text{(Eq. 5.1)}$$

The resulting carbon monoxide (CO) gas readily combined with metals to form carbonyl compounds and, thus, the methane etc. were removed from the atmosphere. The existing high temperatures would have been too hot for liquid water to condense on the surface of the Earth; however, at high elevations, water vapor might have condensed and produced rain in the upper atmosphere that quickly evaporated as it fell lower toward the Earth's surface (Scientific Psychic, 2017).

Earth's primordial atmosphere had a paucity of matter, as compared to that of the Sun, in volatile and mobile elements and compounds. Had it been otherwise, today's atmosphere and hydrosphere would likely be much thicker and denser. Maison (1971) estimated that: (1) the relative hydrogen content on Earth vs. silicon is lower than in the outer space; (2) nitrogen is lower; (3) carbon is lower; and (4) noble gases are lower. Despite the commonality in outer space of such volatile compounds as H_2, He, N_2, H_2O, CO_2, CH_4, NH_3, they are very scarce in the Earth's atmosphere. It has been proposed that such substantial differentiation in the primordial Earth's matter occurred in the pre-planetary stage of the solar system's evolution, when the Sun was at the *t*-Taurus stage of stellar evolution. It may have happened due to removal of the volatile components from the internal portions of the protoplanetary gas-dust cloud into its periphery (into the giant planet formation zone).

The volatile compounds and elements found on the Earth, e.g., H_2O, CO_2, N_2, HCl, HF, HI, etc., probably came in a bonded state from comets, asteroids, etc. Water was likely bonded with hydro-silicates; nitrogen, with nitrides and nitrates; carbon dioxide, with carbonates; halogens, with haloids, etc. Therefore, as the Earth's atmosphere developed, these components were buried in sediments. Some of these residual chemically active volatile compounds were released during the shock explosions of planetesimals hitting the Earth. These residual volatiles were likely then reabsorbed by the ultramafic-regolite on the Earth's surface and buried under new layers of the precipitating meteoritic matter.

Toward the end of the Hadean time, volcanic activity increased the percentage of carbon dioxide in the atmosphere. The Earth's surface changed from molten lava, cooling to a solid rock and liquid water started to accumulate on the Earth's surface.

For the above reasons, Earth's atmosphere, by the end of the Hadean time, was composed mainly of inert nitrogen and noble gases. Due to a short helium *flight time* from the Earth's atmosphere on the order of 10^6 years, its partial pressure in the primordial atmosphere over the time of its formation (about 10^8 years) had enough time to equilibrate, so that its concentration in the atmosphere was similar to that of today. The product of ^{40}K radioactive decay (^{40}Ar) was not likely to have been present in the atmosphere in noticeable amounts. Therefore, it is reasonable to anticipate that the partial pressure of the noble gases in the primordial atmosphere did not exceed 2×10^{-5} atm.

It is much more difficult to estimate the partial pressure of the other active components in the primordial atmosphere, e.g., H_2O, N_2, CO_2, and CO. To calculate them, it would be necessary to know the absorption and reactive capacity of the gases with the ultramafic regolite. The regolite contained free metals, e.g., Fe, Ni, Pt, Co, C, etc., although their partial pressures likely did not exceed 10^{-4} atm.

Earth's Second Atmosphere (Archean time, 4.0 to 2.5 BY ago)

At the beginning of the Archean time, the Earth's crust was cooling. As shown in Figure 5.2, an increase in the Earth's tectonic activity marked the beginning of the Archean and reached maximum levels toward its end. During the Archean time, water vapor in the atmosphere began to condense from the Earth's atmosphere, forming the oceans and decreasing the water

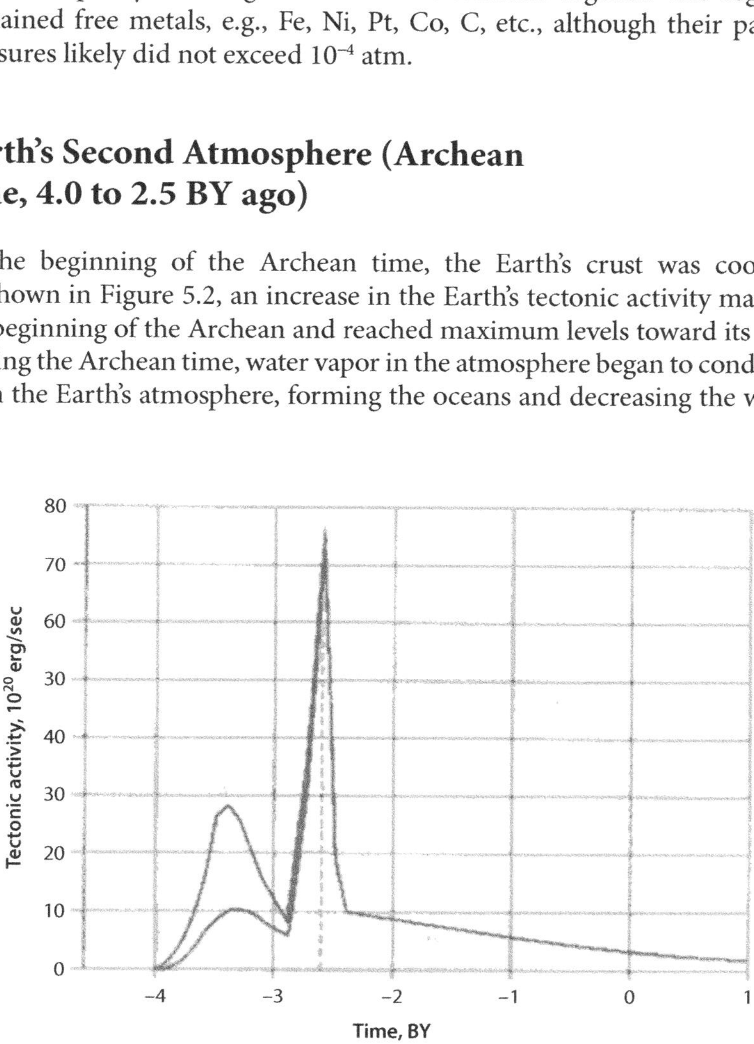

Figure 5.2 The Earth's tectonic activity as measured by depth of heat flow from the mantle. *Curve 1* – average for Earth as a whole. *Curve 2* – tectonic activity within a wide ring belt of the Archaean Earth crust formation above the differentiation zone of Earth's matter. Dashed line corresponds to the time of Earth's core separation.

content in the atmosphere (Figure 5.1). Continuous rainfall on the Earth's surface for millions of years led to the buildup of oceans. Scientific Psychic (2017) suggested that as steam condensed into liquid water on the Earth's surface, the atmospheric pressure of the Earth became lower, permitting water to remain as liquid as the Earth cooled. The liquid water absorbed greater volumes of the volatile gases, e.g., ammonia and methane (as dissolved gas), removing these gases from the atmosphere and by creating ammonium compounds, amines and other nitrogen-containing substances:

$$NH_3 + H_2O \rightarrow NH_4^+ + OH^-. \qquad \text{(Eq. 5.2)}$$

The chemical reaction of water with volatile gases (e.g., sulfur dioxide and carbon dioxide) produced acid rain (e.g., carbonic acid), which, in turn, reacted with other compounds on the Earth's surface forming new minerals:

$$CO_2 + H_2O \rightarrow H_2CO_3 \qquad \text{(Eq. 5.3)}$$

Scientific Psychic (2017) reported that microfossils of sulfur-metabolizing cells have been found in sediments (≈3.4 MY ago) along with the first aquatic photosynthetic organisms that originated about this time. The oxygen produced by *cyanobacteria* (blue-green algae) during the Archean time reacted with the metal ions in the anoxic sea. However, it was not until the Proterozoic time before the photosynthetic microorganisms could eventually produce sufficient oxygen to change the composition of the atmosphere. By the middle of the Archean time, the Earth had cooled enough so that much of the water vapor in the atmosphere had condensed out as liquid water, and there were no clouds. By this time: (1) ammonia and methane were the only minor constituents of the atmosphere (see Figure 5.1); (2) carbon dioxide comprised ≈15% of the atmosphere; and (3) the percentage of nitrogen was ≈75%. Most of the original components of the Earth's atmosphere had either: (1) escaped, (2) precipitated as liquids, or (3) reacted chemically to form solid compounds, which were buried with sediments. By the end of the Archean time the major factors influencing the Earth's atmospheric composition were volcanic activity and the photosynthetic bacteria.

Earth's Third Atmosphere (Proterozoic to mid-Phanerozoic time – 2.5 to 0.54 BY ago)

Monocellular life proliferated during the Proterozoic time developing into multicellular life. During this period, tectonic activity decreased (see

Figure 5.2). Early in the Proterozoic time, anaerobic microbial life thrived on the Earth as there was little oxygen in the atmosphere (see Figure 1.1). As the oxygen content in the atmosphere increased during this period, anaerobic bacteria activity decreased. The various bacteria chemically obtained their energy in several ways. (Scientific Psychic, 2017):

1. Anaerobic methanogens combined hydrogen and carbon dioxide to produce methane and water:

$$CO_2 + 4H_2 \rightarrow CH_4 + 2H_2O \qquad \text{(Eq. 5.4)}$$

2. Anaerobic sulfate-reducing bacteria combined methane and sulfate radicals:

$$CH_4 + SO_4^{-} \rightarrow HCO_3^{-} + HS^{-} + H_2O \qquad \text{(Eq. 5.5)}$$

3. Other organisms, capable of photosynthesis, used the energy of sunlight to convert the abundant carbon dioxide in the atmospheric and water into carbohydrates ($C_6H_{12}O_6$) generating oxygen, which in turn was deadly to the anaerobic bacteria and eventually killed out much of the anaerobic bacteria on the surface of the Earth:

$$6CO_2 + 6H_2O \rightarrow C_6H_{12}O_6 + 6O_2 \qquad \text{(Eq. 5.6)}$$

By the Proterozoic time, the content of the carbon dioxide (CO_2) in the Earth's atmosphere had been nearly depleted and there was only a small percentage of oxygen in the Earth's atmosphere. Nitrogen, N_2, which is chemically inert, became the primary gas in the Earth's atmosphere (see Figure 5.1), as the other gases had become almost completely stripped from the Earth's atmosphere.

Figures 5.1 and 5.3 show that from the beginning of the Proterozoic time to 1.85 BY ago, atmospheric oxygen levels slowly rose as the rate of flora photosynthesis increased. Scientific Psychic (2017) has reported that the shallow seas became partially oxygenated; however, the deeper water layers in the oceans continued to be anoxic. Although photosynthetic organisms had been releasing oxygen since Archean times, the oxygen levels had not built up in the atmosphere because the free oxygen in the atmosphere was being depleted through the oxidation of metals:

$$4Fe + 3O_2 \rightarrow 2Fe_2O_3 \qquad \text{(Eq. 5.7)}$$

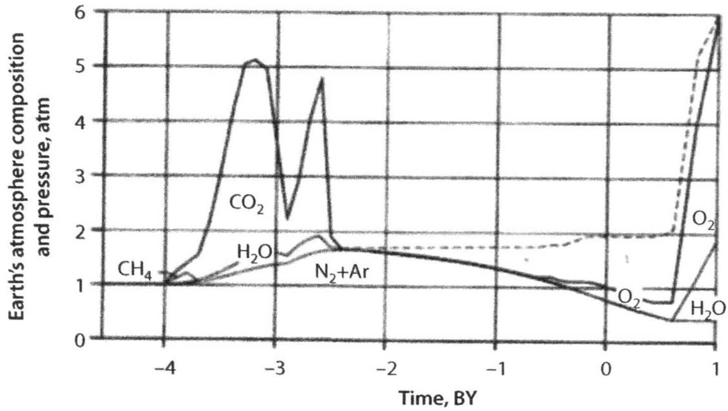

Figure 5.3 Evolution of the composition of the Earth's atmosphere and its pressure with time (dashed line is the atmosphere pressure assuming the absence of bacterial nitrogen consumption).

and by the oxidation of methane:

$$CH_4 + 2O_2 \rightarrow CO_2 + H_2O \qquad \text{(Eq. 5.8)}$$

to yield carbon dioxide and water in the presence of ultraviolet (UV) radiation.

The cooling of the Earth during the Proterozoic time stabilized the land masses and reduced the volcanic outgassing of carbon dioxide (Scientific Psychic, 2017). Some additional oxygen was consumed by oxidation of minerals in the Earth's crust; however, greater quantities of oxygen were generated by plants so that additional free oxygen accumulated in the atmosphere, killing the anaerobic bacteria near the Earth's surface. This created an opportunity for the development of aerobic life forms.

The atmospheric composition stabilized (between 1.85 to 0.85 BY ago) with the Earth's atmosphere containing ≈10% oxygen (Scientific Psychic, 2017). Photosynthetic organisms were still producing oxygen at a high rate, but the reaction of oxygen with the dissolved minerals in the deep oceans and with sediments and rocks on the Earth's surface did not allow atmospheric oxygen levels to increase. By 0.85 BY ago, the minerals in the sea and on land could not bind as much oxygen, and the excess oxygen began to accumulate in the atmosphere.

Today's Atmosphere (Phanerozoic time, 0.542 BY ago to today)

The Cambrian period, at beginning of the Phanerozoic time, is marked by an abundance of multicellular life. This was a time when: (1) most of the major groups of animals first appeared; (2) vegetation covered Earth's surface; and (3) the atmospheric oxygen concentration was ≈30%. By 228 MY ago, the Earth's atmospheric oxygen concentration level was ≈15%, and by the end-Cretaceous (100 MY ago) oxygen had risen to ≈23% (see Figure 5.3). For the last 100 MY years, the percentage of oxygen in the atmosphere has fluctuated between 18% and 23% to the present level of about ≈21%.

The Earth's Future Atmosphere

Figure 5.3 illustrates the Earth's atmospheric composition and the resulting pressure over the entire time of the Earth's existence along with a negative future projection for life on the Earth. This figure shows both the historic and future projections of the atmospheric composition. Sorokhtin and Ushakov (2002) projected a sharp increase in the oxygen partial pressure in 600 MY in the future which is the result of abiogenic oxygen degassing in the process of the Earth's core formation. At present, the formation of the outer layer of the Earth's core occurs because of the reduction of the iron oxide according to the following reaction:

$$2FeO \rightarrow Fe \bullet FeO + O. \qquad \text{(Eq. 5.2)}$$

The released oxygen, under high pressure and release of the compression energy (due to a smaller volume of magnetite molecule) will be bonded with the iron oxides forming the magnetite component of the mantle:

$$3FeO + O \rightarrow Fe_3O_4 + 76.48 \text{ kcal/mole.} \qquad \text{(Eq. 5.3)}$$

After a total oxidation of the mantle silicate iron to magnetite stoichiometry, formation of the Earth's core matter will have to be accompanied by the release of free oxygen not consumed by any reactions:

$$2Fe_3O_4 \rightarrow 3Fe \circ FeO + 5O. \qquad \text{(Eq. 5.4)}$$

As a result, after 600 MY in the future, additional free oxygen will begin entering the Earth's atmosphere. The Earth's atmospheric pressure will rise above 10 atm. This change of atmospheric pressure will increase the greenhouse effect on the Earth (see Eq. 2.26). The Earth's average surface temperatures will rise above 180 °C. At this temperature, the oceans will vaporize, increasing the atmospheric pressure even higher. The Earth's surface temperatures will rise to over 600 °C (which is higher than that on Venus: ≈460 °C). In such an environment, there is little possibility of preservation of life as we know it.

6

Nitrogen in Earth's Atmosphere

Origin of Earth's Atmospheric Nitrogen

Nitrogen, N_2, is a colorless, odorless and tasteless gas that makes up 78.09% (by volume) of the Earth's atmosphere today. It is nonflammable gas and does not support combustion.

UIGI (2017) notes that the density of nitrogen gas is slightly lower than that of air. It has a low solubility in water and is commonly considered and can be used as an inert gas, although it is not truly inert, as it forms nitric oxide and nitrogen dioxide with oxygen, ammonia, hydrogen, and nitrogen sulfide. The nitrogen compounds are formed naturally through biological activity. Nitrogen compounds are also formed at high and moderate temperature with the aid of catalysts. At elevated temperatures, nitrogen combines with active metals, such as lithium, magnesium and titanium to form nitrides. Nitrogen is required for many biological processes, and used as a fertilizer (often in the form of ammonia or ammonia-based compounds).

Nitrogen condenses at its boiling point, −195.8 °C (−320.4 °F), to a colorless liquid that is lighter than water.

	Venus	Earth	Mars
Carbon Dioxide (CO_2)	96.5%	0.03%	95%
Nitrogen (N_2)	3.5%	78%	2.7%
Oxygen (O_2)	Trace	21%	0.13%
Argon (Ar)	0.007%	0.9%	1.6%
Methane(CH_4)	0	0.002%	0

Figure 6.1 Comparison of composition of Venus, Earth and Mars atmospheres. (After Earth Science, 2014.)

Nitrogen is the most dominant gas in today's Earth atmosphere. Figure 6.1, compares the compositions of the atmospheres of Venus, Earth and Mars. The Earth's atmospheric nitrogen concentration is 78%, as compared to 3.5% for Venus and 2.7% for Mars. Reasons for today's apparent lack of nitrogen in Venus's and Mars's atmospheres are that they appear to have had negligible degassing of their mantles (volcanism or plate tectonics) and absence of bacteria. Two major sources controlling the nitrogen content of Earth's atmosphere have been volcanoes (degassing of mantle) and bacterial-fixing of organic material.

Nitrogen is a moderately active element that reacts weakly with inorganic compounds. There are, however, nitrogen-consuming bacteria, that remove nitrogen gas from the atmosphere and circulate it through soil after being dissolved in water, bonding and fixing it to organic matter. The first and most primitive bacteria, *Procaryotes* and *Archaeans,* probably appeared soon after the emergence of life on Earth. It is likely, that in the Archaean time, the primitive bacteria included some nitrogen-consuming bacteria. Organic matter bonded with nitrogen is found in buried oceanic sediments. Also, nitrogen is bonded in nitrates and nitrites fixed by thunderstorm activity. Thus, bacterial life activity and thunderstorms over a long geological evolution have likely stripped nitrogen from the atmosphere and significantly lowered the partial pressure of nitrogen in the Earth's atmosphere.

Today's nitrogen cycle (Figure 6.2) is a description of how nitrogen is converted into various chemical forms as it circulates among the atmosphere,

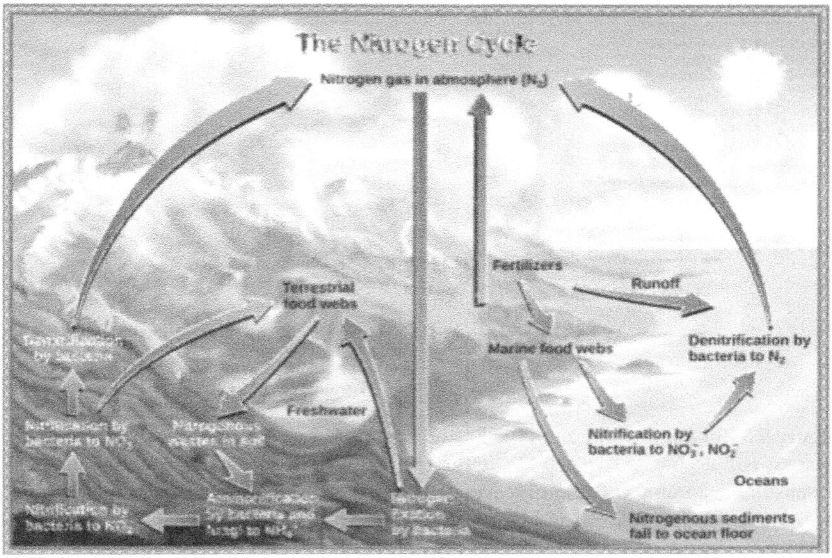

Figure 6.2 Today's nitrogen cycle. (After AP Biology, 2017.)

terrestrial, and marine ecosystems. The bonding of nitrogen is carried in the nitrogen cycle by biological and physical processes (Wikipedia, 2017). Today's cycle of nitrogen is far different from those of earlier time periods, before the development of plants and bacteria. As discussed in this chapter, the Earth's atmospheric pressure, composition and temperature has varied over time, dependent upon the active geological processes.

The nitrogen content in the Earth's atmosphere (Figure 5.1) increased in the Late Archaean time due to introduction of bacteria (biomass) into the oceans (Figures 6.3). This concept is supported by the nitrogen-organic compounds found in the Archaean oceanic sediments. Figure 6.4, *Curve 3*, illustrates the increase in atmospheric partial pressure and atmospheric nitrogen content (see Figure 5.1), as compared to its earlier primordial atmospheric pressure, which was due to degassing from the mantle and loss of other atmospheric volatile gasses by absorption.

During the Proterozoic time, as shown in Figure 5.1, the Earth's atmosphere was almost entirely nitrogen and then the nitrogen content decreased due to an increase in bacterial action which removed nitrogen from the atmosphere. This change in the atmosphere's nitrogen content and lower partial pressure resulted in the cooling of the Earth's surface, which continued through the Proterozoic and Phanerozoic time. Eq. 2.26 demonstrates the relationship between the atmospheric pressure and temperature, the

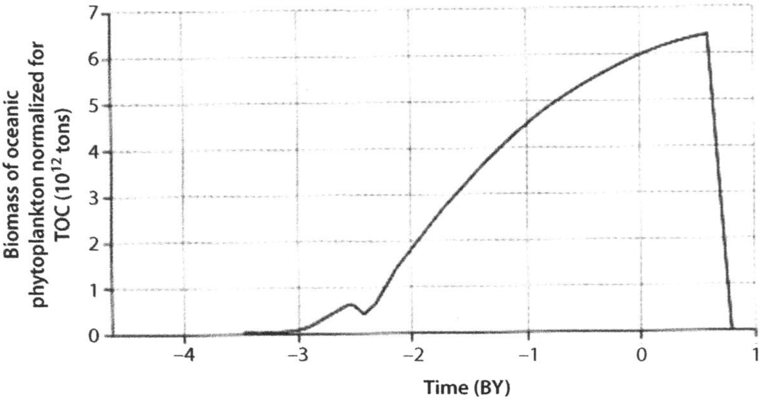

Figure 6.3 Biomass evolution in the Earth's oceans. The authors predict that the biomass will decline to 0 in 600 MY in the future due to the degassing from the mantle in substantial amounts of abiogenic oxygen along with the associated greenhouse effect (Sorokhtin, 1974; Sorokhtin and Ushakov, 2002). (After Sorokhtin *et al.*, 2011, figure 12.7, p. 447.)

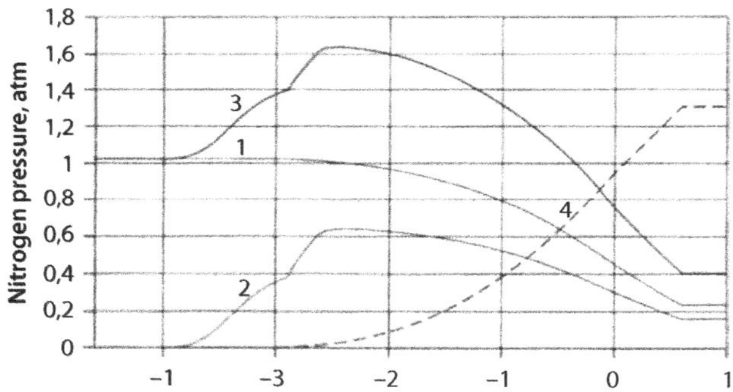

Figure 6.4 Evolution of the Earth's atmospheric nitrogen partial pressure: *Curve 1* – nitrogen of young Earth's primordial atmosphere. *Curve 2* – nitrogen degassed from the mantle. *Curve 3* – pressure of total body of atmospheric nitrogen. Curve 4 – mass of nitrogen removed from atmosphere by nitrogen-consuming bacteria (recalculated using pressure).

Earth's surface temperature declining due to a decline of total atmospheric pressure; decline in the nitrogen's partial pressure is due to a change in nitrogen content of the atmosphere. This total cooling effect on the Earth's surface, over 4.5 BY, has not exceeded 5.4 °C, whereas according to the climatic paradox, the historic cooling over that period reached 70 °C.

This decline in the nitrogen's partial pressure can be attributed to: (1) bacterial metabolism and (2) thunderstorm activity in a humid atmosphere.

The available estimates show that the effect of nitrogen bonding to organic compounds by thunderstorms has a substantially lower effect than that by bacteria. The thunderstorm-bonded nitrogen results in less stable nitrite and nitrate decomposition, which might again release nitrogen into the atmosphere.

On assuming, however, that nitrogen removal from the atmosphere was achieved as well by bacteria as by the thunderstorm activity, the effects are difficult to separate. In calculating the nitrogen absorption effect, one has to consider that the organic nitrogen (N_{org}) accumulated in: (1) the Archean oceanic deposits are continuously removed from the ocean waters, (2) the Archean deposits of the oceanic crust pileup zones, and (3) the plate subduction zones in Proterozoic and Phanerozoic time. Nitrogen was also partially included into the granite-metamorphic rocks of the continental crust or went into the mantle and was partially degassed and released to the atmosphere

The increase of bacterial-bonding of nitrogen-organic compounds and degassing of the mantle in oceanic sediments clearly dominate the nitrogen content of the atmosphere (Figure 6.4, *Curve 3*). Nitrites and nitrates emerging from the thunderstorms are less stable than the nitrogen-organic compounds fixed by bacteria and are likely to be degassed again. Only certain bacteria are capable of consuming and fixing the atmospheric nitrogen to organic matter. These are referred to as nitrogen-assimilating bacteria. Several forms of flora consume nitrogen-organic compounds, e.g., phytoplankton or plants. Flora feed on nitrogen compounds generated by the nitrogen-assimilating bacteria.

Regarding bacterial activity, Wikipedia (2017) presents a partial representation of the nitrogen cycle showing the path of nitrogen: (1) in the atmosphere and (2) soils of the Earth, where a variety of bacterial reactions act on nitrogen gas found in the atmosphere (Figure 6.5).

Estimate of the Earth's Volume of Organic-Nitrogen Sediments

To estimate the volume of bound organic-nitrogen, N_{org}, on the Earth, one can estimate the quantity of organic-nitrogen in the oceanic sediments that has been continuously removed from the oceans through the subduction zones during the Proterozoic and Phanerozoic time. These organic-nitrogen compounds that are partially included are: (1) the granitic-metamorphic rocks of the continental crust or (2) sediments transferred into the mantle. During these processes, nitrogen is partially degassed anew and some re-enters the atmosphere.

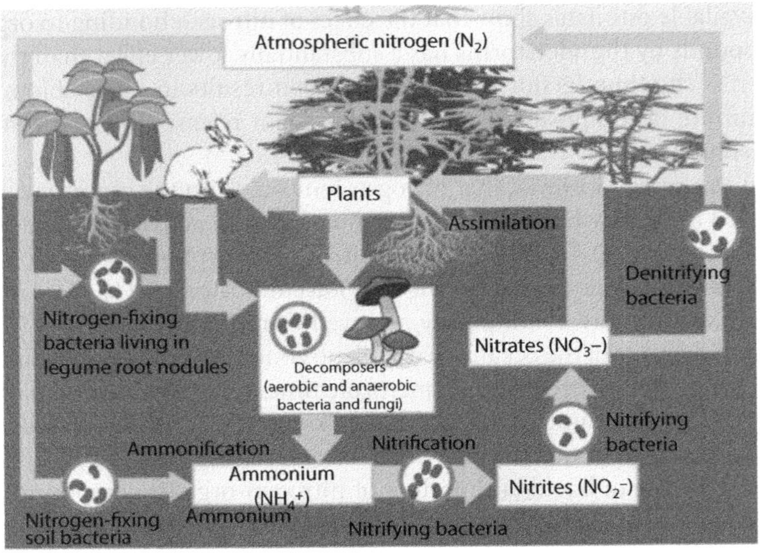

Figure 6.5 Schematic of Nitrogen cycle. (After Wikipedia, 2017), https://upload.wikimedia.org/wikipedia/commons/thumb/f/fe/Nitrogen_Cycle.svg/1200px-Nitrogen_Cycle.svg.png)

According to Ronov and Yaroshevsky (1978) and Ronov (1993), the oceanic deposits (pelagic plus shelf) contain ≈2.7 to 2.86×10^{21} g of C_{org} and in the continental deposits ≈8.09 to 9.2×10^{21} g of C_{org} are buried. Taking into consideration the effect of avalanche-like deposition of solid terrigenous sediments in the nearshore regions of oceans (Lisitsyn, 1984, 1991) with high organic carbon concentration, a slightly higher value of the present-day mass of oceanic sediments and, correspondingly, of the mass of buried organic carbon, C_{org}, is $\approx 3.36 \times 10^{21}$ g. The authors estimate that the mass of nitrogen buried in the: (1) sediments of oceanic floor, and its shelf regions is $\approx 2.35 \times 10^{20}$ g, and (2) continental sediments as $\approx 5.0 \times 10^{20}$ g. This amount of organic-nitrogen was deposited over the average time of existence for the modern oceanic crust, that is, over the recent 5 MY and on the continents, approximately over 400 MY. To determine the organic-nitrogen (N_{org}) accumulation rate during the past geological epochs, one should consider the fact that the element controlling life evolution in the oceans is phosphorus solubility, of which in the oceanic water is limited (Schopf, 1980). The ocean life has evolved mostly within the photo-active water layer. Still, the phosphorus-saturated deeper water reaches this habitable layer through the upwelling zones. It follows that the oceanic biomass has always been approximately proportional to the mass of the ocean *per se* which is shown in Figure 4.11, *Curve 2*. With this data, it is possible to estimate the biomass of the ocean (Figure 6.3).

If the oceanic biomass is proportional to the mass of oceanic water, then one can account also for the approximate mass of the organic-nitrogen (N_{org}) bonded in the deposits and removed together with them during the Earth's geological history from the oceanic sedimento-sphere into the Earth's mantle through the subduction zones of lithospheric plates,

$$(N_{org})_{oc}^{\Sigma} \approx \frac{\dot{N}_{org\,oc}}{m_{oc}^{w}} \int_{t=4\times10^{9}}^{0} m^{w} dt, \qquad \text{(Eq. 6.3)}$$

where $(N_{org})_{oc}^{\Sigma}$ is the total organic-nitrogen mass removed from the oceanic reservoir over the time from $t = -4$ BY ago to the present $t = 0$; $(\dot{N}_{org})_{oc} = (N_{org})_{oc} / \tau$ represents the current (N_{org}) burial rate in the oceanic deposits; $\tau = 120$ MY ago is the average age of today's ocean floor; and (m_{oc}^{w}) and $(m_{oc}^{w})_{o} = 1.42 \times 10^{24}\, g$ are the current value and the present-day water mass in the World Oceans (see Figure 4.11, *Curve 2*).

By numerical integration of Eq. 6.3 shows that during the last 4 BY about 4.3×10^{21}g of nitrogen has been removed from the Earth's atmosphere through marine biota life activity. To this amount must be added the nitrogen preserved in the continental deposits ($N_2 \approx 4.3 \times 10^{21}$ g of organic nitrogen) over a period of close to 400 MY. Thus, the total nitrogen removed from the atmosphere with the small addition of biogenic nitrogen over the life activity on the Earth is $\approx 4.82 \times 10^{21}$g. This would be equivalent to an atmospheric pressure decline of 945 mbar, which is greater than today's nitrogen partial pressure of 765 mbar.

As shown in Figure 5.1, in the Middle Proterozoic time (2 to 0.8 BY ago), the atmosphere was almost pure nitrogen with traces of radiogenic argon. Earth's temperature was much warmer than today and there were no ice sheets anywhere on the continents (Chumakov, 2004b). Therefore, considering the average elevations of the continents at that time, nowhere did the average annual temperature (even at the poles) drop below 0 °C. The adiabatic theory of the *greenhouse effect* (Figure 2.26) requires that for positive temperatures on the surface of the continents, with the average elevations of 1- to 1.7-km, which were formed in the Middle Proterozoic at the poles, the atmospheric pressure should have exceeded 1.2 atm.

The only way one could explain why the nitrogen partial pressure change of about 0 atm in Katarchaean to 1.4 atm in Proterozoic to today's value of 0.755 is by assuming that along with the mantle-degassed nitrogen there was also an effective mechanism for nitrogen removal from the atmosphere (Sorokhtin, 2005a and b, 2008). It can be explained if the present-day atmosphere contains about 40% of the mantle-degassed nitrogen. The nitrogen partial pressure evolution from the mantle is illustrated in *Curve 2,* Figure 6.4.

Chapter 14 discusses the future general decline of the atmospheric pressure and the increase in the climate severity. The resulting temperature increase will affect the general metabolism of the nitrogen-consuming bacteria resulting in a slowdown in nitrogen removal from Earth's atmosphere and destroying the life present now on Earth. It will establish a new equilibrium state of nitrogen content for the atmosphere. This is anticipated in 660 MY in the future.

7

Development of Free Oxygen in Earth's Atmosphere

Oxygen

One fifth of the Earth's atmosphere is made up of oxygen, which is the third most abundant element in the universe by mass. The element oxygen has an atomic number of eight and is the 8th element on the Periodic table. At normal pressure and temperature, the molecules of oxygen can bond to form a dioxide (O_2) which is a colorless gas that composes about 21% of today's atmosphere. Oxygen is the most reactive of the non-metallic elements, a member of the chalcogen group on the periodic table and readily forms compounds with elements other than the inert elements, i.e., helium or neon (Soft schools, 2017).

Today, the movement of oxygen can be seen in the oxygen cycle (Figure 7.1) within its three main reservoirs, the: (1) atmosphere, (2) biosphere (biological matter) and (3) lithosphere (Earth's crust) (CMU, 2003). The form of oxygen in the atmosphere is the gas, free oxygen (O_2). In the other two reservoirs, oxygen is bonded with either organic or inorganic material.

Table 7.1 lists the capacity for storage, the flux in/out and residence time for the three major reservoirs of free oxygen. The lithosphere, shown in

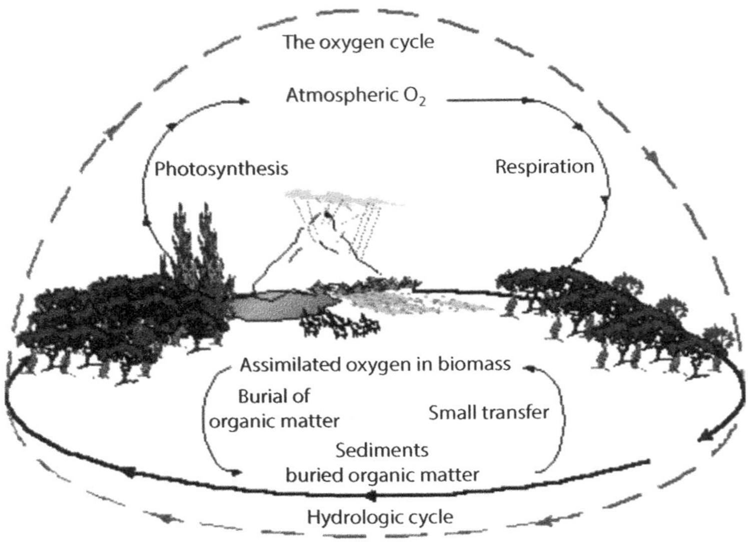

Figure 7.1 Today's oxygen cycle showing three main reservoirs of storage for oxygen: (1) atmosphere, (2) biological, and (3) Earth's crust. (After CMU, 2003.)

Table 7.1 Major reservoirs of oxygen involved in the oxygen cycle. (Data based primarily on estimates from (Walker, J. C. G.; in: Wikipedia, 2017))

Reservoirs	Capacity	Flux in/out, Kg O_2/year	Residence time, years
Atmosphere	1.4×10^{18}	3×10^{14}	4500
Biosphere	1.6×10^{16}	3×10^{14}	50
Lithosphere	2.9×10^{14}	6×10^{14}	500×10^6

Table 7.1, is the largest source of free oxygen. Today, approximately 99.5% of the Earth's oxygen is bound with silicate and oxide minerals. Only a small portion of oxygen is found as free oxygen to the biosphere (0.01%) and atmosphere (0.36%) (Wikipedia, 2017).

History of Free Oxygen in Earth's Atmosphere

As a free gas in the atmosphere, oxygen has only been present for the past 2.3 to 2.4 BY. The development of free oxygen in the Earth's atmosphere first appeared ≈2.5 BY ago (Figure 7.2). The correlation by Govindjee and

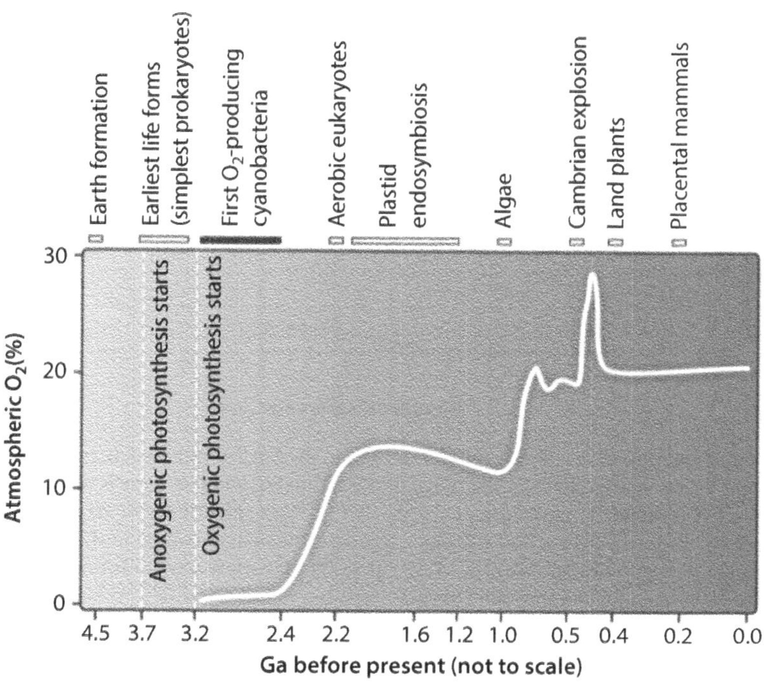

Figure 7.2 Development of free oxygen in the atmosphere. (After Govindjee and Shevela, 2011.)

Shevela (2011) estimated percent free oxygen concentration changes in the atmosphere with the major evolutionary events on the Earth is based on data presented by Falkowski (2006; 2011), Tomitani *et al.* (2006), Kump (2008), Blankenship (2002; 2010), and Hohmann-Marriott and Blankenship (2011). Uncertainties of the events in the evolution of life have been depicted with bars. There are several hypotheses as to when oxygenic photosynthesis was first developed by primitive cyanobacteria-like organisms. Govindjee and Shevela (2011) have summarized the currently available data and depict (by a black bar) that the first cyanobacteria likely evolved as early as 3.2 BY ago or as late as 2.4 BY ago.

The history of free oxygen in Earth's atmosphere can be divided into several phases or stages as shown in Figure 7.3: (1) *Stage 1*—there was little if any free oxygen gas, O_2, in Earth's atmosphere during the Hadean and early Archean times in significant quantities, until the beginning of the Proterozoic time. (2) *Stage 2*—Early in the Proterozoic time there was still little free oxygen in either the atmosphere or dissolved in the oceans; however, the percentage of oxygen increased slightly due to anaerobic bacterial

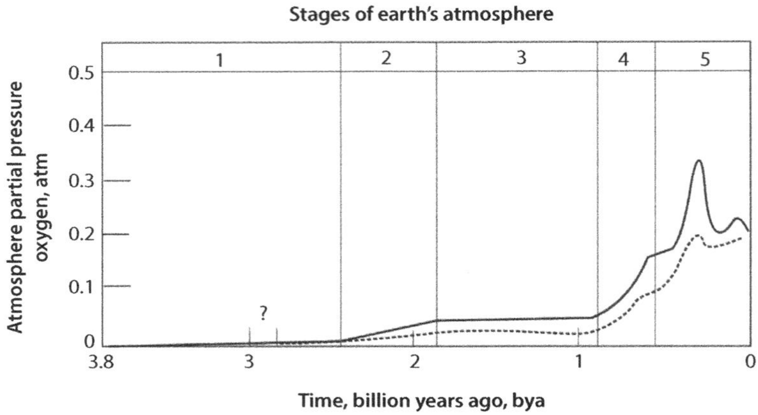

Figure 7.3 Buildup of partial pressure of oxygen in the Earth's atmosphere over time. Solid and dashed lines represent the range of the estimates, whereas time is measured in billions of years ago (BY ago). *Stage-1* – 3.85 – 2.45 BY ago, practically no oxygen in the atmosphere. *Stage 2* – 2.45 – 1.85 BY ago, oxygen produced is absorbed in the oceans and incorporated in seabed rock. *Stage 3* – 1.85 – 0.85 BY ago: more oxygen is produced than can be dissolved in the ocean water (oxygen starts to gas out of the oceans and is absorbed by the land surfaces and enriches the Earth's atmosphere). *Stages* 4 and 5 – 0.85 BY ago – today, oxygen enriches the atmosphere. (Modified after Oxygenation-atm.svg: Heinrich D. Holland derivative work; in: Wikipedia, 2017, https://en.wikipedia.org/wiki/Atmosphere_of_Earth.)

action. (3) *Stage 3*—as a result of the increase of *photosynthetic prokaryotic* organisms in the oceans, bacteria living in the ocean waters produced more free oxygen, which was released as dissolved gas in the water, than could chemically form oxides (Much of this produced free oxygen by the bacteria was removed chemically by oxidation of metals, most notably iron.) Today, these oxidized iron deposits are found in the form of banded iron oceanic sediments on the ocean floor. The excess free oxygen that was not expended in oxidation exceeded the gas solubility of the ocean water and escaped into the Earth's atmosphere. By the mid-Proterozoic time, oxygen began to appear in the Earth's atmosphere in small quantities as shown in Figure 7.3. *Stages 4 and 5,* the development of plants, during the Phanerozoic time, significantly increased the percentage of free oxygen in the atmosphere (first in *Stage 4* by plants within the ocean and later in *Stage 5* when plants developed on dry land). The additional oxygen, produced by bacteria and plants, injected enough free oxygen into the ocean and atmosphere to begin killing off most of the anaerobic bacteria (Wikipedia, 2017).

During the Pre-Cambrian time (Figure 7.3, *Stage 4*), prior to ≈1.9 BY ago, the production of oxygen by the anaerobic bacteria was less than that of the aerobic bacteria and plants. The generation of free oxygen during this era was primarily the result of generation of oxygen by *cyanobacteria* and *microalgae*. In the global oceans, there was an active absorption of oxygen by iron. The concentrations of free oxygen in the Earth's atmosphere were less than 10% of today's concentrations and appeared to have fluctuated greatly. Free oxygen even may have disappeared from the atmosphere again around 1.9 BY ago. The presence of free oxygen in the Earth's atmosphere around 0.541 BY ago when aerobic prevailed over the anaerobic life. New opportunities on the surface of the Earth for the aerobic metabolism made them much more efficient than the anaerobic bacteria in the presence of oxygen. As shown in Figure 7.3, since the start of the Cambrian, atmospheric oxygen concentrations have fluctuated between 15% and 35% of atmospheric volume. (https://en.wikipedia.org/wiki/Geological_history_of_oxygen - cite_note-Berner1999-10) A maximum of 35% was reached towards the end of the Carboniferous (≈300 BY ago).

During the Pre-Cambrian time, the Earth's core separation process iron gradually migrated from the mantle into the Earth's growing core. Simultaneously iron moved into and in contact with the hydrosphere through the oceanic rift zones. There, in the presence of carbon dioxide dissolved in the ocean water, oxygen was distributed throughout the entire ocean in the form of bicarbonate:

$$4Fe + 2H_2O + CO_2 \rightarrow 4FeO + CH_4 + 41.8\ \text{kcal/mole} \qquad \text{(Eq. 7.1)}$$

$$FeO + 2CO_2 + H_2O \rightarrow Fe(HCO_3)_2 \qquad \text{(Eq. 7.2)}$$

Due to the metabolism of some ferro-bacteria species, hematite could have been converted to magnetite (F_3O_4) which later was converted into many of the unique deposits of Pre-Cambrian Jaspillites.

The evolution of the oxygen partial pressure in the Earth's atmosphere over the past 4.5 BY is presented in Figure 7.4. This partial pressure curve of oxygen in Earth's atmosphere is directly related to the types of plant biogenic and bacterial organisms and their oxygen release rate. Figure 7.5 shows the atmospheric free oxygen partial pressure of oxygen for the past 1 BY (Pre-Cambrian to Present). Prior to the appearance of dry land vegetation, the free oxygen partial pressure was proportional with the ocean-water mass. Oxygen absorption had to be also proportional with the iron ore deposits earlier estimated by Sorokhtin (2004); Sorokhtin and Ushakov (2002). Figure 7.6 shows the total rate of iron-ore formation accumulation over the last 4.5 BY.

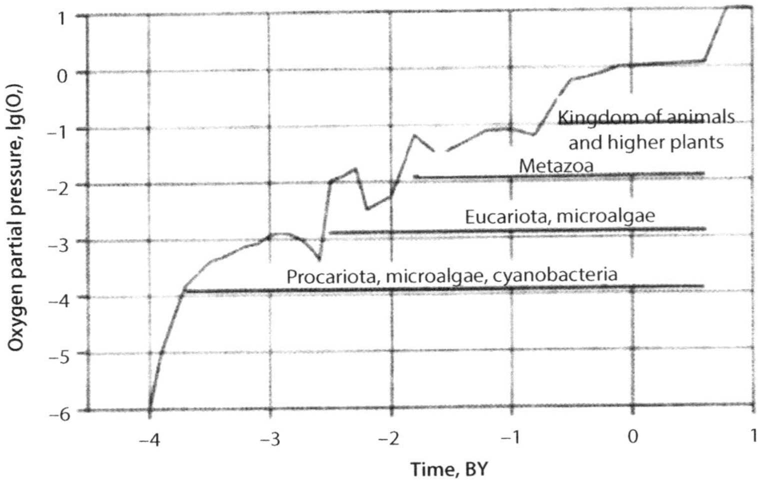

Figure 7.4 Oxygen partial pressure evolution in the Earth's atmosphere (logarithmic scale). In Pre-Cambrian time, oxygen was generated only by the oceanic biota. In the Phanerozoic time additional oxygen was generated by dry–land vegetation. It is assumed that oxygen generation by Archaean prokaryotes (*cyanobacteria*) was by one order of magnitude lower than by Proterozoic *eukaryote* microalgae.

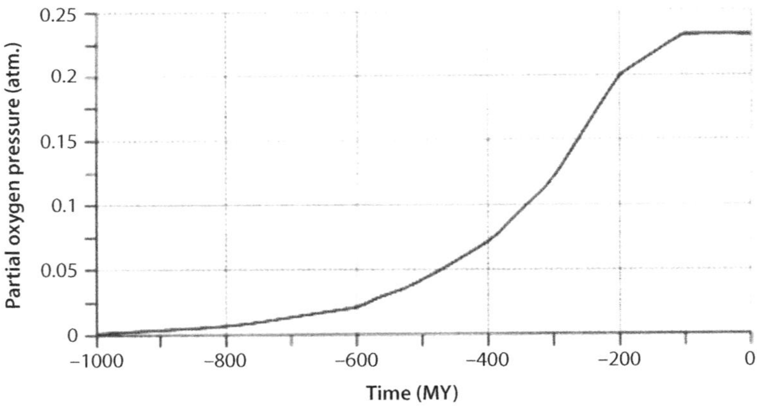

Figure 7.5 Atmospheric free oxygen partial pressure distribution over the last 1 BY (the present-day oxygen partial-pressure is 0.2315 atm.). (After Sorokhtin *et al.*, 2011, figure 12.10, p. 453.)

The biogenic oxygen accumulation in the Earth's atmosphere can be estimated by use of the following relationship:

$$p(O_2) = \frac{aM_{wo}}{b\mathrm{Fe} + c} + p(O_2)_{cont}, \qquad \text{(Eq. 7.3)}$$

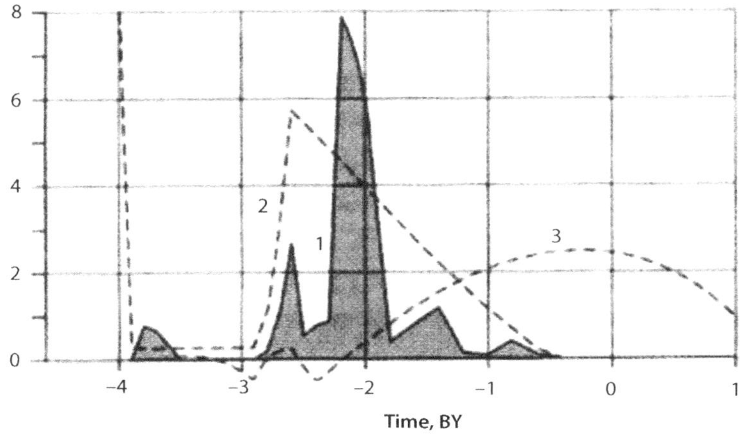

Figure 7.6 Theoretical accumulation rate of the Pre-Cambrian iron-ore formations. *Curve 1* – Total rate of iron ore formation, 10^9 t/year. *Curve 2* – Metallic iron concentration in the convecting mantle, %. *Curve 3* – Ocean surface position relative to the average standing level of mid-oceanic range ridges, km.

where M_{wo} is the water mass in the ocean (see Figure 4.11); Fe is the iron ore accumulation rate (see Figure 7.6); $p(O_2)_{cont}$ is the contribution by dry-land vegetation to the total oxygen partial pressure (in this estimate it was assumed that $p(O_2)_{cont} \approx 0.1$ atm.); factors *a*, *b* and *c* are selected so that the oxygen partial pressure in today's atmosphere would be ≈0.231 atm and for the atmosphere of 600 MY ago would have been lower by one order of the magnitude: ≈0.0231 atm. Using Eq. 7.3 (the logarithmic scale) enables comparison of the major stages in the evolution of Earth's biota as presented in Figure 7.7 (microfossils) and Figure 7.8 (tree of life).

Figure 7.6 demonstrates that every epoch of ferric deposition in the Pre-Cambrian corresponds to an oxygen partial pressure minimum. This can be explained when the oxidized iron is an active oxygen absorber. Also, oxygen partial pressure evolution can be compared with the major stages of the evolution of life on Earth as shown in Figures 7.7 and 7.8. New forms of life on the Earth have emerged following an increase in the atmospheric free oxygen partial pressure, i.e., after the end of the period accumulation of mass iron ore deposits. It should be noted that a change from *thermophilic prokaryote* to more *cryophiles eukaryote* microorganisms likely occurred due to substantial climate cooling in the Early Proterozoic.

As metallic iron migrated from the Pre-Cambrian mantle and migrated toward the Earth's core, biogenic oxygen began accumulating in the Earth's atmosphere as shown in Figure 7.4; this led to a significant restructuring of the entire Earth's biota and emergence of higher life forms (Sorokhtin

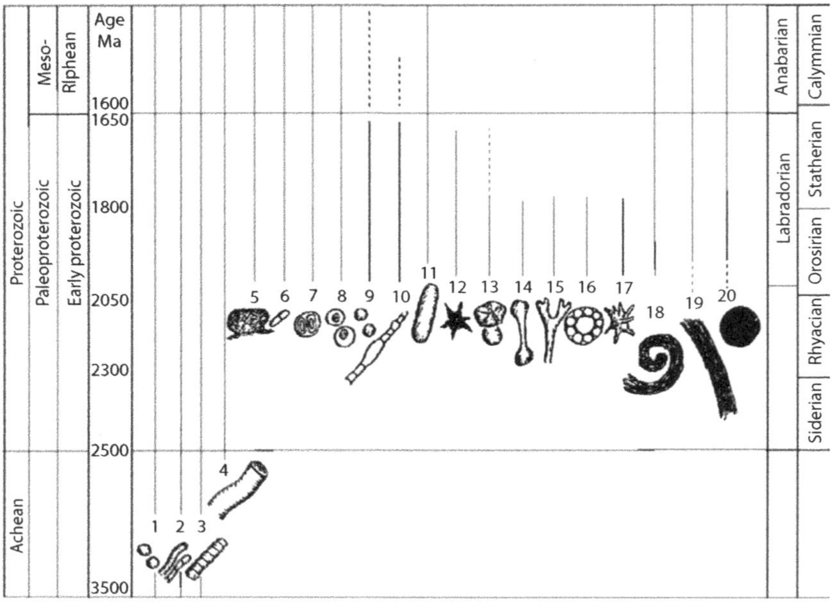

Figure 7.7 Major microfossil remains distribution in the Archaean and Early Proterozoic time (Semikhatov *et al.*, 1999). Common in the Archaean were mostly singular globular and filamentary nanobacteria (1, 2), trichomes and (3) possibly cyanobacteria filaments (4). The diversity of Early Proterozoic microfossils extends from cyanobacteria (4); (5–7), coccoid forms (8,9), trichomes (10) to impressions of large and morphologically complex (11–17), spiral (18), tape-like (19), round and globular (20) forms. (Sorokhtin *et al.*, 2011, figure 15.3, p. 533.)

and Ushakov, 2002). Figure 7.7 illustrates many of the changes in the major microfossil record in the Archaean and Early Proterozoic Eons. The increase in the free oxygen atmospheric partial pressure above 30 mbar was followed literally by an explosion in life's biota variety at the beginning of the eon. This period is often referred to as the *Cambrian Explosion.* It was a time that the skeletal biotic forms and most major types of existing life emerged (Figure 7.8). The emergence and bloom of higher life forms on Earth appears to have been facilitated by a 20 °C decline of planet's average near-surface temperature as supported by Eq. 2.26 and is thus related to a change in the Earth's atmospheric pressure.

The growth in the oxygen partial pressure in Late Riphaean and Phanerozoic (especially in the Mesozoic) was mostly compensated for by the decline, at the same time interval, of nitrogen's partial pressure. After the broad evolution of flowering plants (the major oxygen "manufacturers" at the end Mesozoic), the atmospheric oxygen partial pressure reached

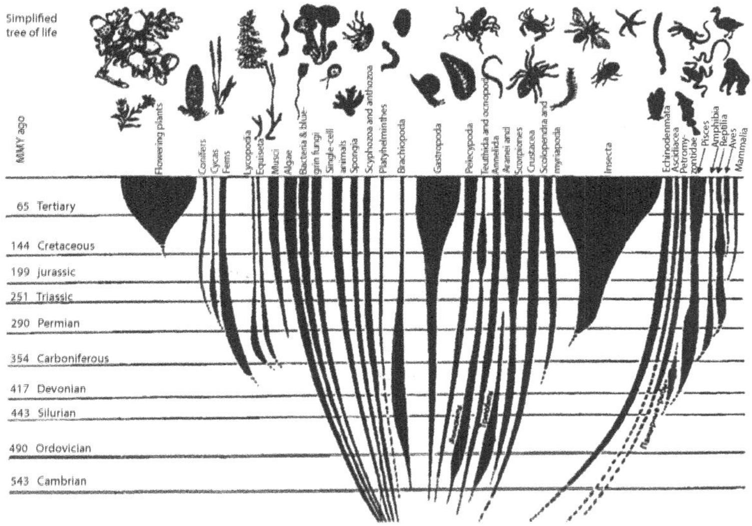

Figure 7.8 "Tree of life". The evolution of life at the Proterozoic/Phanerozoic time boundary was like a biological explosion. (Attenborough, D., *Life on Earth: A Natural History* (1984); in: Sorokhtin *et al.*, 2011, fig. 15-4, p. 534.)

its maximum value and even temporarily exceeded the declining nitrogen partial pressure. However, the growth of the oxygen partial pressure could not be unlimited as its growth was controlled by negative feedback, substantially stabilizing the free oxygen partial-pressure in the Earth's atmosphere. Among the feedback items was the rotting of organic remains and forest fires, which bonded the atmospheric free oxygen to carbon, forming carbon dioxide. Thereafter, upon hydration of the Earth's crustal rocks, were bonded in carbonates under reactions of:

$$\begin{aligned}&2CaAl_2(Si_2O_8)(\mathit{anorihte}) + 4(H_2O) + 2(CO_2) \rightarrow \\ &\quad Al_4\{Si_4O_{10}\}\{OH\}_8\,(\mathit{kaolin}) + 2Ca\{CO\}_3(\mathit{calcite}) \\ &\quad + 110.54 \text{ kcal/mole}\end{aligned} \qquad \text{(Eq. 7.3)}$$

or:

$$\begin{aligned}&4Mg_2(SiO)_4(\mathit{olivine}) + 4(H_2O) + 2(CO_2) \rightarrow \\ &\quad Mg_6\{Si_4O_{10}\}\{OH\}_8(\mathit{serpentine}) + 2Mg(CO_3) \\ &\quad (\mathit{magnesite}) + 72.34 \text{ kcal/mole.}\end{aligned} \qquad \text{(Eq. 7.4)}$$

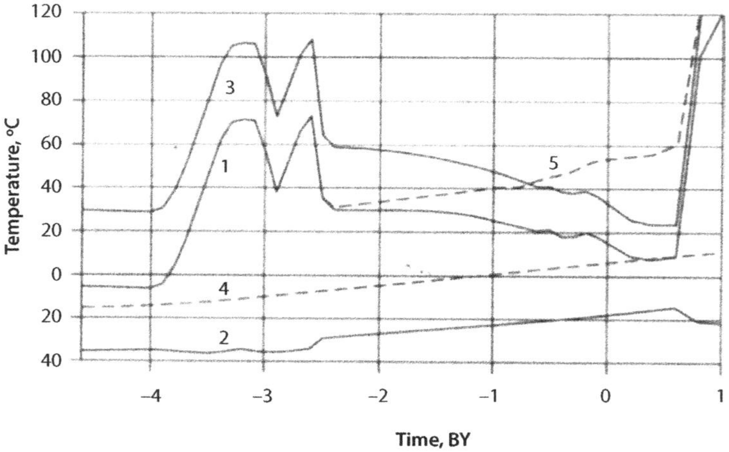

Figure 7.9 Averaged evolution of Earth's climates at a constant Earth's precession angle of $\psi = 24°$. *Curve 1* – Average Earth's surface temperature at sea level estimated using Eq. 2.27. *Curve 2* – Effective Earth's temperature using Eq. 2.25. *Curve 3* – Earth's atmospheric greenhouse effect using Eq. 9.1. *Curve 4* – The absolute black body temperature from Eq. 2.3 at a distance of Earth to the Sun describing the increase, with time, of the Sun's luminosity, *S*. *Curve 5* – Average Earth's temperature assuming that there was no nitrogen consumption by bacteria.

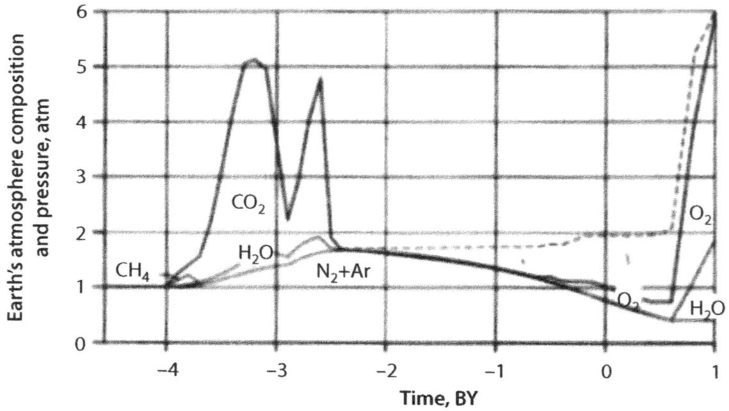

Figure 7.10 Evolution of the Earth's atmospheric composition and pressure evolution with time. The dashed line is the atmospheric pressure in the absence of bacterial consumption of nitrogen.

After reaching a stationary level of oxygen (approximately 230 mbar), nitrogen partial pressure continued to decline, and the atmospheric pressure in Cenozoic began to decline again, resulting in a decline in the atmospheric pressure and a cooling of the climate.

The general decline in atmospheric pressure could not continue without limit, as ≈600 MY in the future, when a drastic increase in the partial pressure of the abiogenic oxygen is anticipated. This problem is the result of Earth's core formation processes shown in Figure 7.9. The suggested increase in near-surface temperature, according to Eq. 2.26, will exceed a temperature of 180 °C resulting in the destruction of all life on the Earth. At this temperature and pressure, there can be no liquid water on the surface of the Earth. The variation of the Earth's atmospheric composition and partial pressures for the life of the planet is shown in Figure 7.10. The only conclusion is that life on Earth is finite and will not continue forever. The time of existence for life on Earth will be ≈4.6 BY, with most of this time (≈4 BY) having already expired with 0.6 BY remaining.

8

Development of Methane in Earth's Atmosphere

Methane the Gas

Methane molecules have a high capacity to absorpt infrared photons. According to estimates of EPA, the GWP (Global Warming Potential) of methane is 21 times higher than the GWP of carbon dioxide over a 100-year time span. Methane released into the atmosphere ascends to the upper layers of the troposphere where it reacts with ozone, ultimately producing (through chemical reactions) water and carbon dioxide:

$$CH_4 + O_3 + \text{solar radiation} \rightarrow CO_2 + H_2O + H_2 \qquad \text{(Eq. 8.1)}$$

Due to its much lighter molecular weight than that of the other components of the Earth's atmosphere at sea level, hydrogen rises into the stratosphere, whereas, because of their heavier molecular weight, water vapor and methane gas can only rise to the border of the troposphere and stratosphere. The gases involved in Eq. 8.1 absorb infrared radiation from the Sun in different portions of the infrared spectrum (see Figure 9.4). The rate of oxidation of methane in the atmosphere depends on the availability of the free OH radicals. The estimated life-time of methane, before it breaks down, in the atmosphere (as estimated by IPCC and adapted by the EPA) is in the range of 8 to 12 years. Thus, the effect of the release of methane gas into the atmosphere, is generating additional amounts of CO_2. For every molecule of CH_4 in Earth's atmosphere, 21 molecules of CO_2 are generated. The effect of this additional release of methane into the atmosphere is the equivalent of an additional release of CO_2 which leads to a cooling of the Earth's atmosphere due to the: (1) increased convection within the troposphere and (2) decreased total atmospheric pressure (see Eq. 2.26). Oxidation of methane is a primary source of water vapor in today's upper troposphere. Therefore, methane gas together with the water vapor, help: (1) shield the Earth's surface from the solar irradiation and (2) lower the average surface temperature of the Earth.

Today's major sources of methane entering the atmosphere are shown in the global methane cycle (Figure 8.1): (1) volcanic activity, (2) marshes and tundra, (3) rice paddies, (4) livestock, (5) hydrocarbons, i.e., production

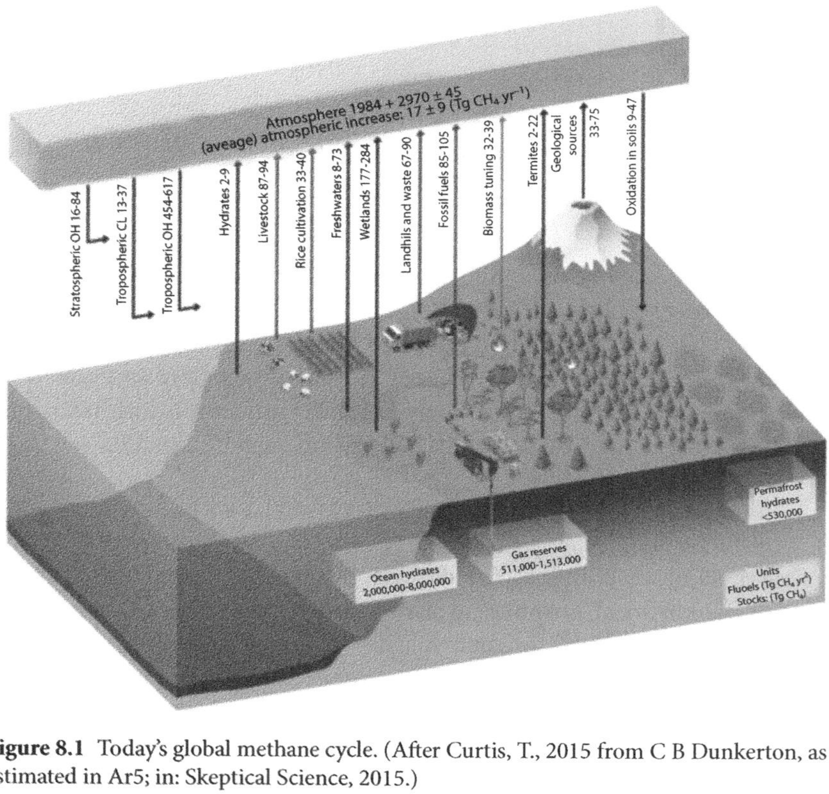

Figure 8.1 Today's global methane cycle. (After Curtis, T., 2015 from C B Dunkerton, as estimated in Ar5; in: Skeptical Science, 2015.)

and burning, (6) natural gas migration to the Earth's surface and (7) landfills and waste. A great volume of methane is released in volcanic eruptions and degassing of the mantle material. Yasamanov (2003) estimated that $\approx 5 \times 10^{15}$ g of CH_4 are released yearly to oceanic water at the spreading zones of mid-ocean ridges. The quantity of methane released to the atmosphere from marshes and tundra is estimated between 5 to 70×10^{13} g annually. From tundra, the amount of CH_4 released is about 4×10^{13} g/year. Enteric fermentation contributes 2 to 20×10^{13} g/year. The total amount of methane released annually into the atmosphere from natural sources is estimated at ≈ 2 to 3×10^{15} g. Methane is also generated because of many human activities. Rice paddies produce about 5×10^{14} g of methane annually. Oil and gas production, burning and related operations add up to 9×10^{14} g of methane annually, which is one order of magnitude lower than the amount of methane released from the natural sources. A great amount of methane is released to the atmosphere because of gas migration to the Earth's surface from coal, oil, and gas deposits and the Earth's mantle (Khilyuk *et al.*, 2000, and Robertson and Chilingar, 2017). The total amount from this latter source is unestimated, although it is likely $\approx 5 \times 10^{15}$ g/year. The burning of fossil fuels contributes 1×10^{13} g/year of methane to the atmosphere.

The content of methane in the atmosphere has been gradually increasing over the last century. Although the methane content constitutes only about 1.8 ppm in the Earth's atmosphere, national and international policy makers have declared the methane gas extremely dangerous to the Earth's climate because of a high potency (about 100 times more potent than CO_2 over the time span of 20 years) of its molecules to absorb the infrared radiation. Together with the growing contents of CO_2 and other greenhouse gases, it is supposedly capable of causing drastic changes in the Earth's climate. According to the conventional anthropogenic theory of global warming, due to the absorption of the infrared radiation by the molecules of the atmosphere, these molecules intercept infrared photons in the lower layer of troposphere, thus warming the Earth. This anthropogenic theory is the basis for strong political and economic actions against further expansion of fracking, e.g., used in the shale-gas production. Additional discussion about the effect of increasing the carbon dioxide content of the Earth's atmosphere shows that increasing the carbon dioxide content will actually lower the Earth's temperature and not increase it as many people erroneously believe (see Chapter 10).

Again, this conventional anthropogenic theory (backed and promoted by IPCC and other national and international organizations over the last 25 years) has been based, without scientific support, solely on the phenomena of heat transfer in the troposphere as only occurring by radiation, whereas in reality, it is convection and not radiation that is the overwhelming form of heat transfer in the troposphere as documented in Chapter 10. In the troposphere today, heat transfer occurs 67% by convection, 8% by radiation, and 25% by water vapor condensation (Sorokhtin *et al.*, 2007). Moreover, analyzing the postulates in the incorrect conventional theory, it completely ignores the fact that many molecules of the *greenhouse gases* (H_2O, for example) intercept the infrared solar irradiation in the upper layers of stratosphere and not in the troposphere where the methane is located.

Historic Levels of Methane in the Earth's Atmosphere

Methane and ammonia are important trace gases in the Earth's atmosphere. Methane and ammonia were present at the initial stages of the Earth's development (see Figure 5.1). The methane content in the Earth's atmosphere today is 0.00017% (1.7 parts per million by volume). As one of the greenhouse gases of the Earth's atmosphere, methane traps a significant amount of IV radiation. Today's quantity of methane in the atmosphere is the result of a balance between production on the Earth's surface and destruction in the atmosphere. The source of the Earth's methane is: (1) when organic matter decomposes in oxygen-poor environments, e.g., marshes, rice paddies, or the digestive systems of herbivores; (2) from combustion, e.g., burning, of hydrocarbons; and (3) from inorganic reactions (consequently termed abiogenic methane).

Early in the Earth's history, about 3.5 BY ago, there was 1,000 times as much methane in the atmosphere as there is today (see Figure 5.1). The earliest methane that was released into the atmosphere appears to be the result of volcanic activity. The first bacteria contributed methane to the atmosphere by converting hydrogen and carbon dioxide into methane and water.

The emergence of the Precambrian iron ore formations is based on the iron-oxidizing processes by thermal dissociation of the CO_2-saturated oceanic waters and hydration by the same waters of iron-containing oceanic crust rocks. In these processes, methane was also generated:

$$\begin{aligned} &4Fe + 2H_2O + CO_2 \rightarrow 4FeO + CH_4 + 41.8 \text{ kcal/mol} \\ &FeO + 2CO_2 + H_2O \rightarrow Fe(HCO_3)_2 \end{aligned} \qquad \text{(Eq. 8.2)}$$

Today, the mantle does not contain free iron. The methane generation from the oceanic rift zones historically occurred during the hydration of the oceanic crustal rocks containing bivalent iron, mostly according to the following reactions:

$$4Fe_2SiO_4(\text{fayalite}) + 12Mg_2SiO_4(\text{forsterite}) + 18H_2O + CO_2 \rightarrow$$
$$4Mg_6(SI_4O_{10})(OH)_8(\text{serpentine}) + 4Fe_2O_3(\text{hematite})$$
$$+ CH_4 + 183.4\ \text{kcal/mol.} \qquad \text{(Eq. 8.3)}$$

Due to isotope fractioning between the carbon dioxide and methane, methane was enriched in the light isotope ^{18}C, whereas the heavy isotope ^{13}C mostly migrated to carbon dioxide. Such isotope shifts occur, for instance, in the hydrotherms of the *black smokers* which arise on serpentinites. Methane in such sources usually has $\delta^{13}C \approx -13$ to $-14‰$, whereas the oceanic water-dissolved HCO_3^- and CO_2 have the shifts close to $\delta^{13}C = -5.5‰$ (Sorokhtin *et al.*, 2001). This indicates that the isotopic exchange between carbon dioxide and methane occurs according to the following reaction:

$$^{12}CO_2 + {}^{13}CH_4 \rightarrow {}^{13}CO_2 + {}^{12}CH_4 + 0.412\ \text{kcal/g} \qquad \text{(Eq. 8.4)}$$

in the direction of lowering $\delta^{13}C$ in methane.

As the iron ore formations were highly developed in Precambrian, it is expected that the iron released from the mantle to the hydrosphere, as well as the hydration of the iron silicates occurred as described by Eq. 8.3. This was followed by profuse abiogenic methane generation.

The maximum iron transport to the ocean was accompanied by a maximum methane generation rate. This iron oxide release, in turn, resulted in increasing the mass of the methane-consuming bacteria. The iron oxide release may be described as follows:

$$Fe + H_2O \rightarrow FeO + H_2 + 5.84\ \text{kcal/mol} \qquad \text{(Eq. 8.5)}$$

Fractioning of carbon isotopes always produces a lighter methane isotopic composition, and, hence, a lighter carbon composition in organic matter C_{org} in the bacteria which consumed this methane. This is a likely explanation of the extremely low shifts from $-50‰$ to $\approx -80\ \delta^{13}C_{org}$ in the methane-consuming bacteria. It may also explain local minima in $\delta^{13}C_{org}$ distribution at the time when the largest mass of the iron ore formations was deposited at the end of the Archaean and in the Early Proterozoic time.

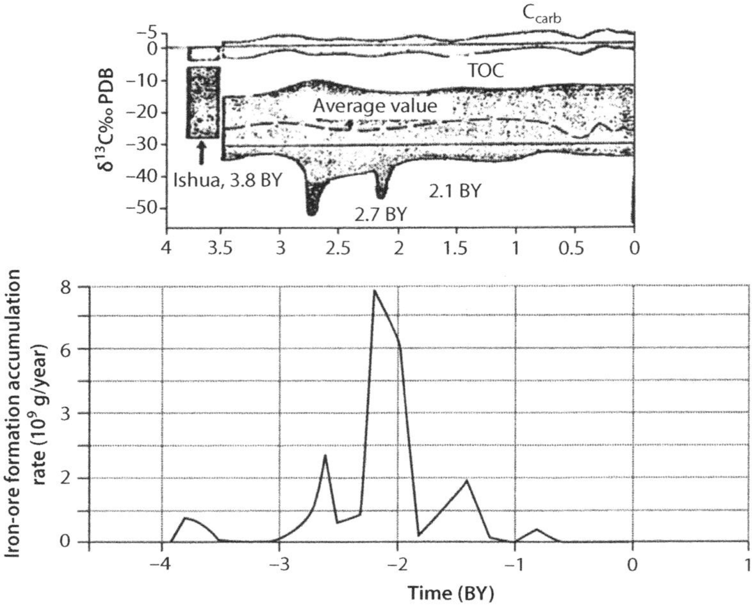

Figure 8.2 Carbon isotopic shifts in organic matter versus epochs of the Precambrian iron-ore formations. Upper diagram: $\delta^{13}C_{org}$ and $\delta^{13}C_{carb}$ distributions in the Earth's evolution (Schidlowski, 1987). Lower diagram. The two most outstanding Precambrian iron-ore formation deposition epochs clearly correspond with the two local minima on the envelop of the minimum $\delta^{13}C_{org}$ values. Unfortunately, the upper diagram does not include their complementary positive $\delta^{13}C_{carb}$ anomalies which reached +8 to +11‰ PDB in the Early Proterozoic time (Semikhatov *et al.*, 1999).

When examining the carbon isotopic shift in organic matter distribution, one can observe the epochs of the massive iron ore deposition on the ocean floor corresponding with a minimum in the organic carbon $\delta^{13}C_{org}$ shift. However, despite a smaller intensity of jaspelite formation rate in the Late Archaean, the $\delta^{13}C_{org}$ isotopic minimum amplitude was the largest at that time.

Every epoch of massive iron ore sedimentation on the ocean floor corresponds to a minimum rate in the Late Archaean time, the $\delta^{13}C_{org}$ isotopic minimum amplitude was the largest. The reason for this may be the existence in the Archaean time, of a high percentage of carbon dioxide in the atmosphere, whereas in the Early Proterozoic time the CO_2 partial pressure had declined, and the rate of methane generation had decreased.

The time iron ore accumulations in the Late Archaean and Early Proterozoic time also corresponds to the two periods of broad stromatolite development (Figure 8.3). Most likely, the nutrition base for the

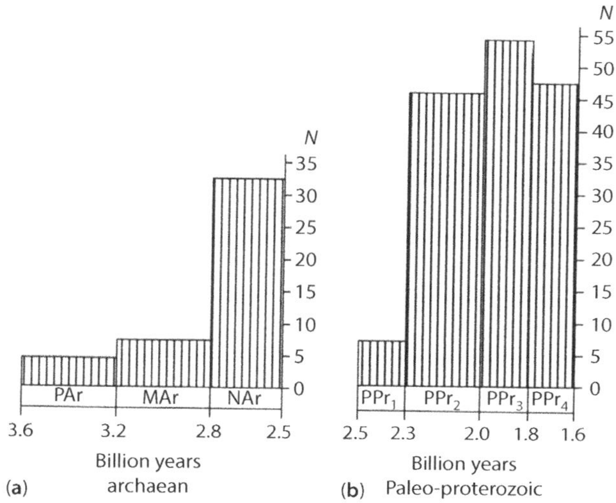

Figure 8.3 Changing of the number of stromatolithic formations in Archaean (a) and Proterozoic (b). (After Semikhatov et al., 1999.) N represents the number of formations with stromatoliths. The Archaean stromatoliths are much smaller in mass than the Early Proterozoic ones ("Paleoproterozoic" by Semikhatov). (After Sorokhtin *et al.*, 2011, p. 455, figure 12.12.)

stromatolites was the abiogenic methane generated in the metallic iron oxidation process according to the reaction of Eq. 8.2 and hydration of the iron silicates according to the reactions like Eq. 8.3.

The abiogenic methane generation rate $\dot{m}(CH_4)$ tends to be controlled by the oceanic crustal formation rate (Dmitriyev *et al.*, 2000; Sorokhtin *et al.*, 2001) equal to $V(H_b + H_{sp})\rho$, where V is the oceanic crustal area increase rate in the rift zones (see Figure 8.4, *Curve 3*); H_b and H_{sp} are, respectively, the oceanic crust's basalt-gabbroid and serpentinite layer thickness; and $\rho \approx 2.9$ g/cm^3. During the epochs of the mass iron ore formation deposition, additional methane release occurred at a rate proportionate to the iron ($\dot{F}$) ore accumulation rate in these formations.

In the Precambrian rift zones, the methane generation occurred due to the hydration of iron silicates under reactions like Eq. 8.3 as well as due to the participation of the free (metallic) iron under reaction Eq. 8.2. Both reactions proceeded within the oceanic crust upon hydration of its rocks by the oceanic water. There is a possibility that similar reactions occurred during the Archaean time because the saturation of the primordial regolite with rain water was saturated with carbon dioxide, as the regolite at that time covered a significant portion of the Earth's surface (see Figures 8.5B, 8.5 C and 8.6).

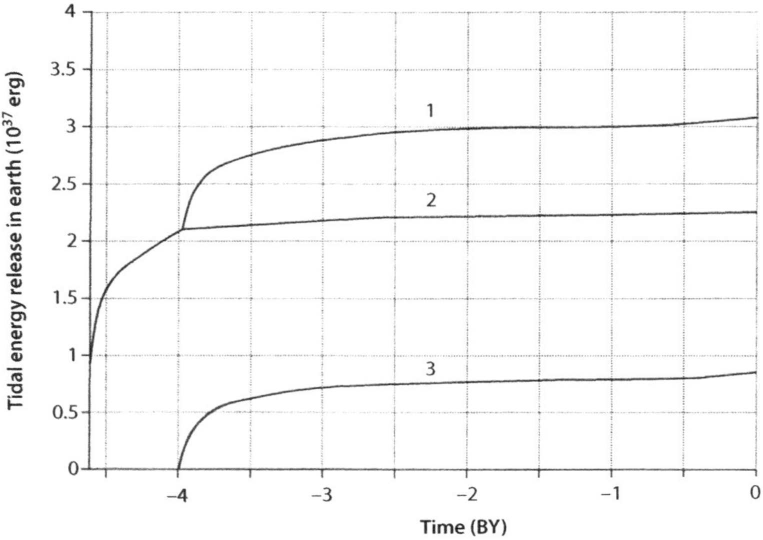

Figure 8.4 Tidal energy release: *Curve 1* – in Earth; *Curve 2* – in mantle; and *Curve 3* – in hydrosphere. (After Sorokhtin *et al.*,2011, fig. 5-10, p. 178.)

Normalizing the calculation results to a single scale can be done by normalizing all methane generation reactions to their unit mass. For example, using the iron oxidization reaction (Eq. 8.2) one mole of methane (16 g) within the rift zones and primordial regolite, forms because of the combination of 304 g of the required reagents to generate this reaction. Thus, the scale factor for this reaction is: $k = 16/304 = 0.0526$. Similarly, the oceanic crust iron-containing rock and regolite hydration reactions are:

$$\begin{aligned} &4Mg_2(SiO_4)(\text{forsetrite}) + 4H_2O + 2(CO_2) \rightarrow \\ &\quad Mg_6(Si_4O_{10})(OH)_8(\text{serpentine}) + 2Mg(CO_3) + 65.05\ \text{kcal/mol} \\ &2CaAl_2Si_2O_8(\text{amorite}) + 4H_2O + 2(CO_2) \rightarrow \\ &\quad A_{14}(Si_4O_{10})(OH)_8(\text{kaolin}) + 2Ca(CO_3)(\text{calcite}) + 110.54\ \text{kcal/mol.} \end{aligned} \quad \text{(Eq. 8.6)}$$

has a scale correction factor, $n = 16/2876 = 0.00556$. The abiogenic methane generation rate, with the excess of CO_2, may be presented as:

$$\frac{d(CH_4)}{dt} = 0.8 \times 10^{15} U \rho n C_{FeO} (Hsp + 0.12 H_b) \left(\frac{S_O}{S_G} \right) + 0.1\, k \left(\frac{d(Fe)}{dt} \right) \quad \text{(Eq. 8.7)}$$

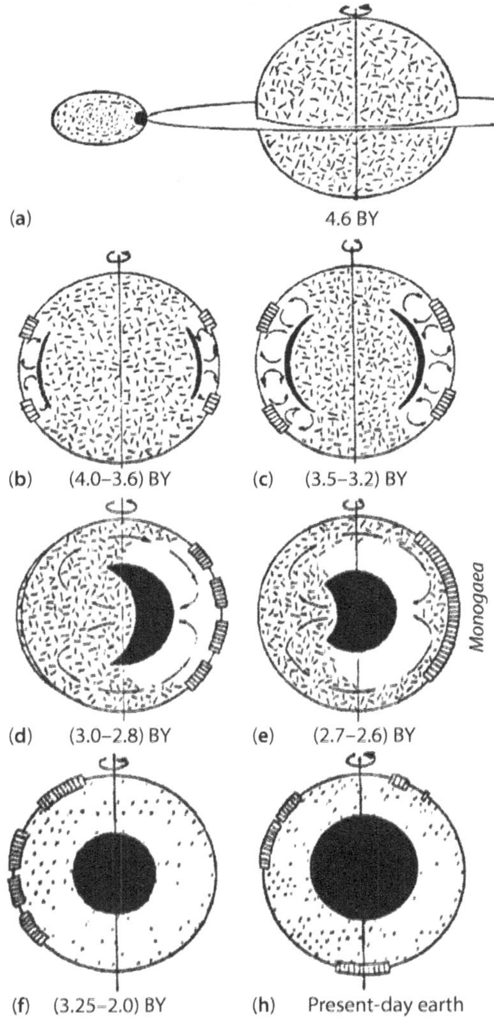

Figure 8.5 Earth structure evolution: (a) the young Earth and moon formation; (b–f) consecutive stages of the Earth's core separation and formation; (g) present-day Earth. The dashes represent the primordial matter; solid black is iron and its oxides melts; white is the Archaean time depleted mantle impoverished in iron, its oxides, and siderophilic elements; the dots are the current-type mantle; and the boxes are continental massifs. (After Sorokhtin *et al.*, 2011, fig. 4.1, p. 117.)

where S_O is the oceanic crustal area (see Figure 8.6); $S_G = 5.1 \times 10^{18}\ cm^2$ is the Earth's surface area; C_{FeO} is the ion-oxide concentration in the mantle matter and the oceanic crust (see Figure 8.7); factors 0.1 and 0.8 relate to the recovery extent of a given component ($FeO_2 \cdot FeO$ from the oceanic

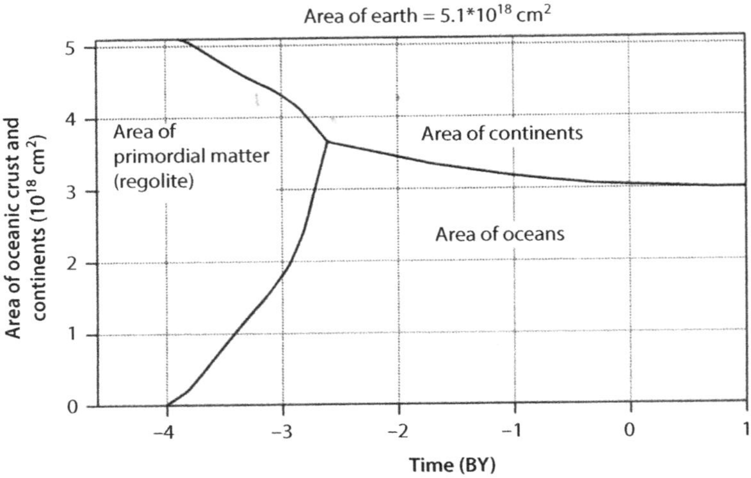

Figure 8.6 Evolution of the areas of the oceans and continents with time. (Sorokhtin *et al.*, 2011, figure 7.4, p. 245.)

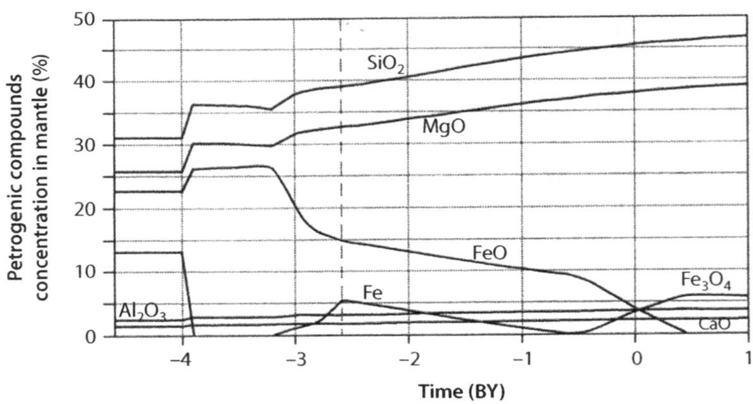

Figure 8.7 Evolution of major petrogenic elements and compounds in the convecting mantle. (After Sorokhtin *et al.*, 2011, figure 4.16, p. 145.)

crust layers). Using Eq. 8.3, the methane generation in today's oceanic crust is 1.86×10^{12} g/year. This is close to the values obtained by Dmitriyev *et al.* (2000) of $\approx 2 \times 10^6$ t/year (or 2×10^{12} g/year). The calculated results can be compared with organic carbon isotopic shifts in Figure 8.8. The Precambrian iron ore deposition and methane accumulation clearly correspond to the two minima on the envelope of the minimum $\delta^{13}C_{org}$ values.

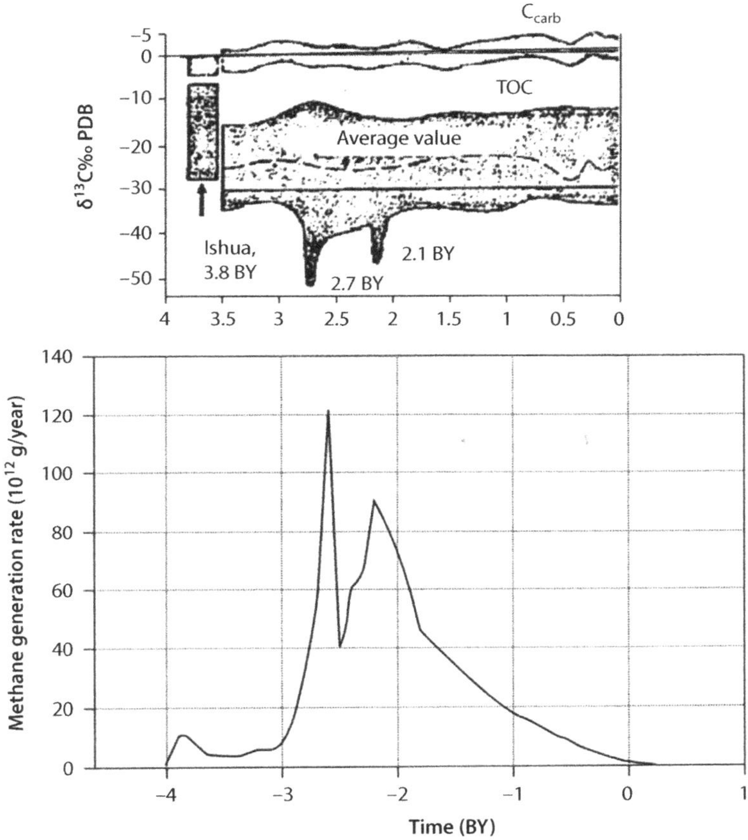

Figure 8.8 Abiogenic methane generation in the oceanic crust plotted using Eq. 8.7 compared with the carbon isotopic shifts in organic matter. Upper diagram: $\delta^{13}C_{org}$ and $\delta^{13}C_{carb}$ distributions versus time (Schidlowski, 1987); lower diagram: abiogenic methane generation rate. Two most outstanding Precambrian iron-ore formation deposition and methane accumulation epochs clearly correspond with the two local minima on the envelop of the minimum $\delta^{13}C_{org}$ values. (After Sorokhtin *et al.*, 2011, figure 12.11, p. 456.)

As shown in Figure 8.8, the generation of abiogenic methane in the oceanic crust ≈4 BY ago, reached 10 million tons per year, and then rapidly declined to a few million tons per year in the Mid-Archaean time. Later, due to the beginning of the Earth's core separation (see Figure 8.5D and 8.5E) and a significant increase in the tectonic activity (see Figure 5.2), the generation of methane (see Figure 8.8) significantly increased and by the end of the Archaean time reached maximum value of about 120 million tons per year. During the Archaean and Proterozoic time, tectonic activity diminished significantly along with methane generation (≈40 million tons per year). Massive deposition of iron ore formations began in the Early Proterozoic (≈2.2 BY ago) as indicated (Fe) in Figure 8.7. This resulted in renewed intensification of methane generation of ≈90 million tons per year. Subsequently, the abiogenic methane formation gradually decreased to today's rate of 1.8 million t/year. This process will stop in ≈600 million years, when, due to the Earth's core separation process, the mantle becomes totally devoid of bivalent iron.

There is a strong correlation between the organic carbon isotopic shift and methane generation rate. As a result, there are massive Early Precambrian iron ore formations. This is logical as iron oxidizing under the reaction of Eq. 8.2, and its further accumulation in iron ore formations, is accompanied by generation of the abiogenic methane.

Inasmuch as methane is the nutrition base for the cyanobacteria and methane-consuming bacteria, it left noticeable traces in the geological record. Due to carbon isotope fractionation in the exchange reactions between carbon dioxide and methane according to the following reaction:

$$^{12}CO_2 + {}^{13}CH_4 \rightarrow {}^{13}CO_2 + {}^{12}CH_4 + 0.412 \text{ kcal/g}(CH_2) \qquad \text{(Eq. 8.8)}$$

This shows that methane is always enriched in the light carbon isotope. The remaining carbon dioxide later migrates to the carbonates to became isotopically heavier. For that reason, organic matter formed from methane-consuming bacteria will always have a lighter carbon isotope composition, sometimes as low as $\delta^{13}C_{org} \approx -50\,^{o}/_{oo}$ and lower.

This phenomenon may have been the cause of the local $\delta^{13}C_{org}$ distribution minima exactly at the times when the iron ore formations were deposited at the end of the Archaean time and Early Proterozoic time at the greatest rate.

Accelerated methane generation during the deposition of the massive iron ore formations must have affected the evolution intensity of stromatolites (Figure 8.3), which are the remains of laminated bacterial mates (films). If so, a noticeable fraction of these films was formed by methane-consuming bacteria. The bloom of these life forms must have occurred during the massive deposition of the iron ore formations and profuse methane generation. Semikhatov *et al.* (1999) indicated that this time interval is marked by a burst in the growth of stromatolite buildups (see Figures 8.3 and 8.8).

A portion of the abiogenic methane could have concentrated in accumulations of oceanic deposits of gas-hydrates. Massive shungite deposition occurred in Karelia during the Late Jatulian time (the second half of Early

Proterozoic time). The shungites are coaly matter that are likely graphitized remains of previous hydrocarbon accumulations.

Beside the generation of methane in the oceanic rift zones, a nonrecurring abiogenic methane release into the atmosphere occurred at the beginning of Archaean time due to the hydration of the ultramafic regolith still covering most of the planet (see Figures 8.5B, 8.5C and 8.6). The methane release intensity may be estimated by assuming that over a period of 100 MY, from ≈4.0 to 3.9 BY ago, the Earth's entire primordial layer of regolith was soaked with rainwater containing carbon dioxide. The initial iron concentration of the regolith was ≈13%, and its oxides, ≈23%. In this situation, with the excess of CO_2 in the water, the original atmospheric methane pressure may be determined as:

$$p_{(CH_4)} = p_r h_r (0.13k + 0.23n)\left(\frac{S_r}{10^3 S_G}\right) \quad \text{(Eq. 8.9)}$$

where the regolith density (by analogy with the lunar regolith) is $p_r \approx 2$ g/cm^2; h_r is the unknown thickness of the rainwater-soaked regolith layer; Sr is the area covered by the regolith (see Figure 8.6); $S_G = 5.1 \times 10^{18}$ cm^2 is the Earth's area. In the methane formation reactions (Eq. 8.2 and Eq. 8.3), the limiting factor is the mass of the degassed carbon dioxide. The authors estimate that by the time 3.9 BY ago, $m(CO_2) \approx 6.4 \times 10^{20}$ g of carbon dioxide was degassed from the mantle. With the assumption that all this gas was spent for methane generation, its mass would then be $m(CH_4) \approx m(CO_2) \times 16/44 \approx 2.33 \times 10^{20}$ g, and its partial pressure would not exceed $p_{(CH4)} \approx 46$ mbar. This methane pressure calculated using Eq. 8.9 corresponds to a carbon dioxide rainwater-soaked regolith layer of 37.5-m thick. The $p_{(CH4)}$ estimate appears to be maximal. The actual methane partial pressure at the beginning of Archaean time was lower, but most likely no lower than 20 mbar.

Methane could not accumulate in the sea basin water in the locations where it was generated (within the primordial regolith and outside the sediment sequences). Thus, it was released directly into the atmosphere making the Earth's atmosphere very reducing. Therefore, the atmosphere at the beginning of the Early Archaean time was significantly reducing and the atmosphere's composition was carbon dioxide-nitrogen-methane. Subsequently, in about 100 MY after the total hydration of the primordial regolith, the Archaean atmosphere's reducing potential must have declined due to methane photodissociation by solar radiation with the formation of formaldehyde:

$$CH_4 + CO_2 + h\nu \rightarrow 2HCOH(\text{formaldehyde}). \quad \text{(Eq. 8.10)}$$

Besides, being affected by the solar ultraviolet (UV) radiation, methane was also oxidized in the humid and warm Early Archaean atmosphere according to the following reaction:

$$CH_4 + H_2O + h\nu \rightarrow CO + 3H_2, \qquad \text{(Eq. 8.11)}$$

with the hydrogen evaporating. Therefore, the Earth's Archaean atmosphere became a carbon dioxide-nitrogen one, with a minor methane content and at the equilibrium humidity content.

As Galimov (2001) has shown, for life to have emerged abiogenically, some simple source organic compounds had to have first accumulated. One such compound may have been formaldehyde, which as methane formed during the iron oxidation:

$$2Fe + H_2O + CO_2 \rightarrow 2FeO + HCOH(\text{formaldehyde}) + 3.05\ \text{kcal/mol}. \qquad \text{(Eq. 8.12)}$$

Beside the formaldehyde, because of the thunderstorm discharges in the Earth's Early Archaean reducing atmosphere, with its nitrogen and methane composition, another *link* in the chain of primary life, the hydrogen cyanide, must have emerged:

$$N_2 + 2CH_4 + Q \rightarrow 2HCN(\text{hydrogen cyanide}) + 3H_2 \qquad \text{(Eq. 8.13)}$$

where $Q = 26.6$ kcal/ mol is the part of the thunderstorm discharge energy absorbed by the Eq. 8.13 reaction. Early in the Archaean time (≈3.9 BY ago) bacteria probably had not yet emerged. If so, the isotopic shifts in organic matter at that time were determined only by the properties of the reactions like Eq. 8.2 and 8.3 with no addition of isotopic shifts due to the bacterial metabolism. Apparently, these oldest Earth's rocks were populated by bacteria only somewhat later and with no participation of the abiogenic methane. That may be one reason why the carbon isotopic ratios about 3.9 to 3.8 BY ago in the Shidlovsky diagram have slightly elevated $\delta^{13}C$ values when compared to subsequent epochs (see Figures 8.2 and 8.8).

The evolution of abiogenic methane partial pressure in the atmosphere was developed by the authors using Eqs. 8.1 and 8.8 with results are presented in Figure 8.9. To make the characterization of the methane generation on Earth more complete, Eq. 8.8 must be amended by: (1) the addition of swamp methane in northern Eurasia and Canada, (2) methane generated

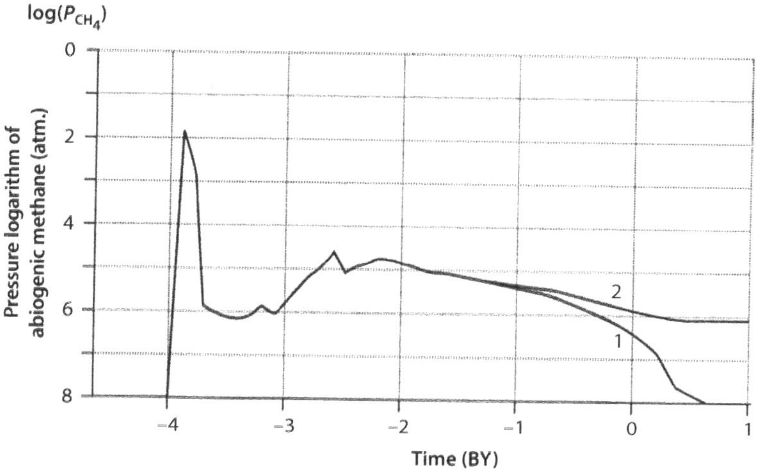

Figure 8.9 Atmospheric methane partial pressure versus time: *Curve 1* – methane generation in the oceanic crust and *Curve 2* – total methane generation in the oceanic crust and on the continents. (After Sorokhtin *et al.*, 2011, figure 12.14, p. 459.)

by soil-bacteria in the humid tropical areas, and (3) methane seeping up from the continental hydrocarbon accumulations.

It is assumed that the process of evolution of partial pressure of methane began about 1.8 BY ago. Today, Earth's atmospheric methane partial pressure is $p_0 \approx 1.216 \times 10^{-6}$ bar ($\log p_0 \approx -5.92$). Thus, one can calculate the total methane partial pressure in the Earth's atmosphere. The authors estimate is that today's methane generation on the continents is 3.3 times that generated by the oceanic crust. The methane partial pressure has fluctuated during the Earth's evolution by at least four orders of magnitude (Figure 8.9).

Methane is an unstable hydrocarbon gas and under normal circumstances rapidly decomposes. For example, the dissolved gas from the oceanic water: (1) escapes into the atmosphere; (2) rises to the stratosphere; (3) rapidly decomposes when it encounters hard solar radiation and breaks down into carbon dioxide; and (4) remains within the atmosphere as carbon dioxide:

$$CH_4 + 2H_2O + hv \rightarrow CO_2 + 4H_2 \quad \text{(Eq. 8.14)}$$

$$CH_4 + 2OH + hv \rightarrow CO_2 + 3H_2 \quad \text{(Eq. 8.15)}$$

Within the oceanic water, methane is consumed by bacteria and then released as a carbon dioxide gas byproduct. Also, when the bacteria die, their organic material breaks down into heat, carbon dioxide and water:

$$CH_4 + \text{bacteria} \rightarrow CO_2 + (\text{OM}). \qquad \text{(Eq. 8.16)}$$

In the Early Precambrian active metamorphism zones, the direct reduction of methane to carbon occurred with the formation of coaly matter (combustible shales, shungite and even graphite). In the late Precambrian, especially in the Phanerozoic time, direct methane oxidation occurred:

$$CH_4 + 4O_2 \rightarrow 2H_2O + CO_2 + 131.2\ \text{kcal/mol}. \qquad \text{(Eq. 8.17)}$$

Despite all these losses, the methane remaining in the sedimentary sequences, especially in the absence of oxygen, could form hydrocarbon accumulations, including gas hydrates. More complex hydrocarbons were also formed from the methane. Lein and colleagues (1998, 2000) believe that the *black smoker* sulfide ores are a unique target for studying genesis of organic matter. The chemical peculiarity of hydrothermal systems which include dissolved H_2, H_2S, CH_4, NH_4, Fe^{2+}, Mn_2^+, CO, CO_2, the NaCl electrolyte, and sulfide minerals as a solid phase (electrodes) gives rise to a possibility of organic compound bacterial synthesis on the sulfide mineral matrix in this giant electrochemical cell. Lein *et al.* (1998) studied lipid fractions in hydrothermal springs using single-dimensional and bi-dimensional chromatography and noted that all similar samples included a broad range of neutral lipids and phospholipids, and that in the water-methanol medium, proteins and DNAs.

Mass-spectrometric analysis of the lipid fractions from the organic matter in sulfide ores confirms the presence of biogenic components, including cellular material of microorganisms. Fatty acids from the present-day sulfide ores are compositionally close to lipids from the biomass of chemoautotroph bacteria within the hydrothermal community in the active fields of mid-ocean ridges (Lein *et al.*, 1998; Lein and Sagalevich, 2000; Perisypkin *et al.*, 1998).

Monitoring of Methane Gas Emissions

Rossenfoss (2015), in an excellent article, discussed the need for better monitoring of methane gas emissions. According to the EPA, the natural gas emissions would rise by 25% by 2025 without further action.

According to the EPA, since 1990, the industries' natural gas emissions are down by 16%. Rossenfoss (2015) stated that this was mainly due to: (1) reduced emissions from controlled devices, (2) installation of plunger lift systems, and (3) green well-completion methods, which capture gas that once escaped from the fluids, flowing back after fracturing.

Based on the study done by the consulting firm ICF (commissioned by EDF), the added cost, using available technology, of a 40% reduction in emissions from natural gas production processing and delivery system would be about 1 cent per MCF produced.

Rossenfoss (2015) showed that great strides have been made by oil/gas companies in the field of emission control. The conclusion was that operators should regularly monitor and maintain their gas-producing wells and equipment.

9

The Effect of the Greenhouse Gases

The Greenhouse Gases

Today, the Earth's atmosphere is primarily composed of nitrogen, oxygen, and argon, which constitute the primary gases of its atmosphere. Other gases present in the Earth's atmosphere are often referred to as trace gases, among which are the group of gases recognized as the *greenhouse gases.* At the Earth's surface, the atmosphere also includes a variety of trace amounts of human introduced chemical compounds. Also many substances of natural origin may be present in miniscule amounts, e.g., as aerosol in an unfiltered air sample, including dust of mineral and organic composition, pollen and spores, sea spray, and volcanic ash. In some industrial areas, various industrial pollutants also may be present as gases or aerosol, such as chlorine (elemental or in compounds), fluorine compounds and elemental mercury vapor. Sulfur compounds, e.g., hydrogen sulfide (H_2S), and sulfur dioxide (SO_2), which may be derived from natural or industrial sources.

The term *greenhouse gas*, GHG, used in this book, is defined as a gas in the Earth's atmosphere that absorbs and emits radiation within the thermal infrared range. This process of absorbing energy from the Sun is the fundamental effect of what is referred to as the *greenhouse effect*. The primary greenhouse gases in Earth's atmosphere are water vapor, carbon dioxide, methane, nitrous oxide, chlorofluorocarbons (CFCs) and hydrofluorocarbons (incl. HCFCs and HFCs). Some people also include ozone as a greenhouse gas; however, due to its significance, it has been separately discussed in Chapter 11.

The Classic *Greenhouse Effect*

The *greenhouse effect*, ΔT is the difference between an average planet's surface temperature, T_s, and its effective surface temperature, T_e, at which this planet can be seen from outer space:

$$\Delta T = T_s - T_e \quad \text{(Eq. 9.1)}$$

As average surface temperature for Earth is approximately 288 K or +15 °C, and for Earth under the classical variant where the precession angle, $\psi = 0°$, $T_e = 255$ K or −18 °C, the greenhouse effect on Earth is currently recognized as 33 °C. However, in consideration of Earth's current precession angle, $\psi = 23.44°$ and $T_e = 263.5$ K (or −9.5 °C) the greenhouse effect is much lower, $-\Delta T = 24.5$ °C.

Figure 9.1 illustrates how energy from the Sun is reflected and/or absorbed and then radiated from the Earth's surface and atmosphere. The absorption of energy from the Sun's light waves by the greenhouse gas molecules, lengthens the light wave as the energy is absorbed by the gas, which is then emitted from the gas molecule with a longer wave length. In general, shorter electromagnetic waves carry more energy than longer ones. A portion of this energy is absorbed by the greenhouse gas molecules in the Earth's lower atmosphere, increasing the rate of vibration (temperature) of molecules. This process of the greenhouse molecules absorption of energy, increasing the vibration (temperature) of the molecules, is referred to as the *greenhouse effect*.

When the concentration and/or pressure of the greenhouse gases in the atmosphere change, the amount of energy that can be absorbed by the atmosphere also changes. However, changing the concentration of a gas in the atmosphere can also change the partial pressure of that gas component, which contributes to the total atmospheric pressure. Eq. 2.26 establishes that changes in pressure affect the absorption of energy by that

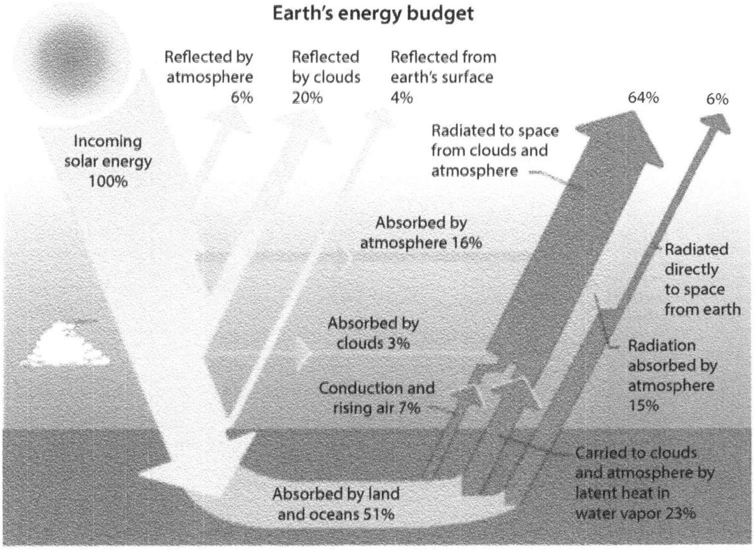

Figure 9.1 Schematic illustration showing the distribution of energy carrying light waves from the Sun to the Earth. (After Lightle, 2008.)

gas. In summary, increasing the gas concentration of carbon dioxide, can result in less absorption of energy if the atmospheric pressure remains the same. Figure 9.2 demonstrates how additional cloud cover can decrease the Earth's surface temperatures by reflecting more (albedo) energy carrying wave away from the Earth's surface. An excellent example of this albedo effect occurs as a result of volcanism where ash is injected into the atmosphere, reflecting more energy from the Earth, resulting in the cooling of the Earth.

The Earth's radiation budget is a concept that Lightle (2008) presented which helps us understand the quantity of energy that the Earth receives from the Sun, and how much of that energy the Earth absorbs and then radiates back to outer space (see Figure 9.1):

The Greenhouse Gases

The dominant atmospheric gases that absorb energy include carbon dioxide (CO_2), methane (CH_4), nitrous oxide (NO_2), methane (CH_4), water vapor (H_2O), ozone (O_3) and any fluorocarbons. The details of what wavelengths each gas can absorb is shown in Figure 9.3. Also shown in this figure is the ability of the whole atmosphere (all gases) to absorb specific wave

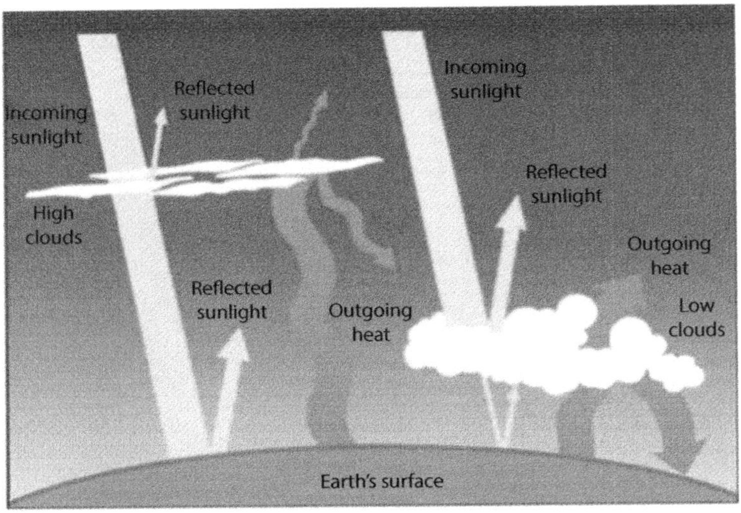

Figure 9.2 Schematic of how clouds affect cooling and warming of the Earth. (After NASA, 2017.)

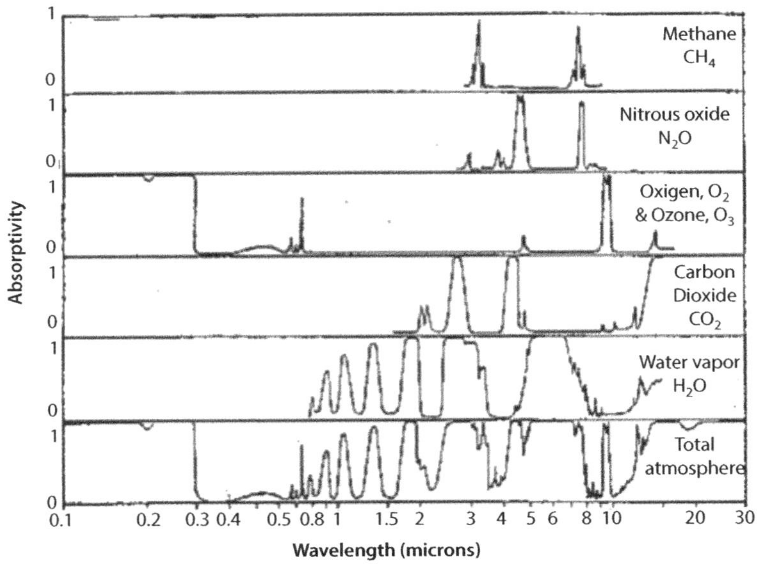

Figure 9.3 Wave length radiation absorptivity for various gases of the atmosphere. An absorptivity of zero means no absorption, while a value of one means complete absorption. The dominant gas absorbers for infrared radiation are water vapor and carbon dioxide, oxygen and ozone. (After J. N. Howard, 1959: Proc. I.R.E. 47, 1459; and R. M. Goody and G.D. Robinson, 1951: Quart. H. Roy. Meteorol. Soc. 77, (153); in: http://www.meteor.iastate.edu/gccourse/forcing/images.html.)

lengths. The dominant absorbers of infrared radiation are water-vapor, carbon dioxide, oxygen and ozone. These gases in the atmosphere are not pollutants as identified by some, but are natural and without them life on the Earth would not be possible.

Some of the radiated heat released by greenhouse gases reaches the Earth's surface, along with heat from the Sun that has penetrated the atmosphere. Both the solar and radiated energy are absorbed by the Earth's surface and its atmosphere and reemitted at longer wave lengths; some of this energy from waves reflected from the Earth's surface are again reabsorbed by greenhouse gases in the atmosphere, perpetuating the cycle. It is thought by some that the greater the concentration of these gases will always result in more heat being absorbed by the Earth and, consequently, the warmer the Earth's surface. Eq. 2.26 demonstrates that if the total atmospheric pressure increases, more heat will be absorbed by the atmosphere and, likewise, if the total atmospheric pressure decreases less heat will be absorbed. Changing the atmospheric composition and keeping the total pressure the same by the addition of a gas with a lower partial-pressure will lower the Earth's ability to absorb energy or its temperature.

According to many scientists, without the greenhouse effect, which smooths out the atmospheric temperature extremes, the Earth's surface temperature would be 5 °F rather than 59 °F. The greenhouse effect has been going on, throughout the Earth's history for the past 4.5 BY. The French mathematician, Joseph Fourier, first recognized the greenhouse effect in 1824 (Mester, 1996). At the end of the nineteenth century, Arrhenius (1896) presented a hypothesis on heating of the atmosphere by increasing the content of carbon dioxide, ignoring the effect it would have on the total atmospheric pressure. Even though it is incorrect, for a long time his hypothesis has been accepted as factual. Unfortunately, it has been incorrectly used to predict disastrous effects by increasing the absorption of heat in the atmosphere, e.g., carbon dioxide. This never verified concept is still believed by many today, even though it has been shown to be wrong (Greenhouse Effect, 1989; Green Peace Report, 1993; Budyco, 1997).

Along the same line, some environmentalists have erroneously predicted that increasing the concentration of greenhouse gases will accelerate climatic change, earthquakes, heavier storms, etc., and that the global surface temperature could rise from 1 to 4.5 °F over the next 50 years, and 2.2 to10 °F in 100 years, e.g., Kyoto Protocol (1997), without any creditable scientific proof. They predict that this proposed rise in temperature would increase evaporation, which, in turn, will increase the global precipitation and more intense rainstorms around the world would occur. These proposed elevated global temperatures would increase melting of Arctic and

Antarctic ice, resulting in a rise of the Earth's sea levels over 0.6 m, flooding the coastal areas over the next 50 years. All these disastrous climatic changes have been attributed to increasing the greenhouse gas concentration in the atmosphere without scientific support or understanding of thermodynamics. Therefore, over the past 50 years, their dire climatic predictions have never happened. Yet today we see many individuals making similar unfounded claims, blaming any recent climatic problems on the increase of anthropogenic carbon dioxide emissions without verification. Serious analyses and critique of these fallacious environmental predictions have been made, based upon geologic history and scientific analysis, by Robinson *et al.* (1998), Sorokhtin (2001a, b), and Khilyuk and Chilingar (2003, 2004, 2006).

An examination of many of these theories shows a lack of understanding of thermodynamics. Some of these numerical calculations and several climatic predictions have often been based on intuitive models, using poorly defined parameters (Greenhouse Effect, 1989). Examination of these environmental models reveals inherent uncertainties in the model parameters (several models contained at least 30 such parameters) making the numerical solution of the problem itself questionable.

Our investigation of the greenhouse effect is based on the scientific relationships of the physical parameters, including the description of the mass and heat transfer within the atmosphere, i.e., synergetic approach (Haken, 1980, 1983; Prigozhin and Stengers, 2003). Many of these future predictions of the Earth's temperature, ignoring the thermodynamics of the atmosphere over the past 40 years, have proven to be incorrect. Today, globally there are polar ice caps and shrinking and growing glaciers, even though several of these earlier predictions forecast that they all would be melted by today.

Understanding the Greenhouse Effect

The Earth's atmosphere is an example of an open dissipative system, which can be described by nonlinear equations of mathematical physics. This permits one to assume that the organization of physical fields and the formation of stable thermodynamic structures can be defined for the Earth's atmosphere, considering the time-space scale as determined by the parameters of the phenomenon. One may use significant and reliable parameters of the atmosphere and the primary characteristics of the processes driving the Earth's climate to construct a climatic model, e.g., one may use such general parameters as the total mass of atmosphere, its thermal capacity, and the averaged energy of solar irradiation. In addition, one must take into consideration, the existence of a strong negative feedback between the spherical *albedo* (albedo: the fraction of solar radiation that is reflected into space) of Earth and its averaged surface temperature. In such an approach, however, the localized details in the description of the greenhouse effect are lost, because the constructed model becomes one-dimensional and averaged over the entire Earth.

In many cases, however, this general approach possesses definite advantages, as one obtains an analytical and unambiguous solution of global problems of the entire planet, e.g., the influence of the composition of atmosphere on the total magnitude of the greenhouse effect. In addition, a constructed general model can be specified, by adding additional parameters and local variables. In the general model, one can incorporate the crucial factors that can further improve the model, e.g., (1) the latitude of a certain locality; (2) inclination of the axis of Earth's rotation to the ecliptic plane; (3) the precession of Earth's and Sun's axis; (4) inflow of additional heat with the air flows (cyclones); (5) the reflection capacity of snow cover, and (6) etc. In such a way, one can construct a three-dimensional or even four-dimensional model (the fourth dimension is the time) of the greenhouse effect. The major factor determining the climate of the Earth is the quantity of solar radiation absorbed by Earth's atmosphere. Also to be considered is the Earth's atmosphere, composition, pressure, and thermal capacity (Sorokhtin, 1990, 2001a, b; Sorokhtin and Ushakov, 2002).

The Greenhouse Effect

The average surface temperature for Earth is ≈288 °K or 15 °C, and its effective radiation temperature is determined by the classic Stefan-Boltzmann law:

$$T_c^4 = \frac{(1-A)S}{4\sigma}, \qquad \text{(Eq, 9.2)}$$

where $\sigma = 5.67 \times 10^{-5}$ erg/cm$^2 \cdot$ s $\cdot$ °C^4 is the Stefan-Boltzmann constant; S, is the solar constant at the distance of the Earth from the Sun ($S = 1.367 \times 10^6$ erg/cm$^2 \cdot$ s); A is the albedo, which is mostly due to Earth's cloud cover ($A \approx 0.3$). According to Eq. 9.2, the effective temperature, T_e, is equal to 255 K (or 18 °C). Therefore, the present-day greenhouse effect for the Earth would be equal to 33 °C.

Applying the Stefan-Boltzmann law to evaluate the heat transfer in the atmosphere: (1) the heat transfer by radiation dominates only in the upper diffuse layers of the atmosphere, e.g., stratosphere, mesosphere, and thermosphere; (2) the heat transfer in the troposphere, the lowest and denser layer, occurs mostly by convection (Sorokhtin, 2001a, b). In the Earth's atmosphere, where the air pressure exceeds 0.2 atm, the heat transfer by convection is dominant. When the temperature of a given mass of air increases, its volume also increases proportionally. This is the model of a system that defines the Earth's climatic atmospheric system, but it is also one that recognizes cooling and heating by convection in the troposphere, rather than a model dependent upon radiation.

A significant conclusion from this observation is that the temperature distribution in the Earth's troposphere must be close to adiabatic, as when the air mass warms and compresses while moving down and cools while expanding and rising. This does not necessarily imply that at any instant, distribution of temperature must be totally adiabatic. There are several processes of heat transfer occurring in the Earth's atmosphere: radiation, convection and conduction.

The adiabatic temperature distribution is determined by the atmospheric pressure, *p*, and by the effective thermal capacity of atmosphere:

$$\frac{pV}{T} = \text{constant (ideal gas law)}, \qquad \text{(Eq. 9.3)}$$

as the volume never changes. The thermal capacity considers its additional heating capability as a result of the absorption of infrared radiation of the Earth's surface by the greenhouse gases and heat of condensation of water vapor in the troposphere. For an adiabatic process, the temperature of gas is a function of pressure that can be presented in the following form (Landau and Lifshits, 1979):

$$T = Cp^{\alpha}, \qquad \text{(Eq. 9.4)}$$

where C is a constant, which can be estimated using the experimental data, $\alpha = (\gamma - 1)/\gamma$; $\gamma = c_p/c_v$ and c_p and c_v are the specific heats of gas at constant pressure and constant volume, respectively. For triatomic gases, CO_2 and H_2O, $\gamma = 1.3$ and $a = 0.2308$, whereas for the diatomic gases, N_2 and O_2, $\gamma = 1.4$ and $a = 0.2857$. Because of condensation of water vapor m troposphere, which process emits heat, the adiabatic exponent, a, decreases, e.g., the average value of this parameter for the humid and heat-absorbing troposphere, $a = 0.1905$, whereas for the dry air, $a = 0.2846$ (Sorokhtin and Ushakov, 1999).

Condensation of water vapor in the troposphere creates cloudiness, which is the primary factor that determines the Earth's albedo, *A*. This process gives rise to a strong negative feedback between the near-surface and radiation temperatures for the Earth, stabilizing the temperature regime of the troposphere. A rising surface temperature increases the water evaporation and the resulting cloudiness in the troposphere, which increases the albedo of planet and the reflection capacity of Earth's atmosphere. The portion of solar radiation reflected to space increases, decreasing the heat (energy) supplied to the Earth. As a result, the average temperature of Earth's surface decreases.

In any system, a negative feedback leads to a linear dependence of the system's output on the system's input. This is a universal property of systems with negative feedbacks. This property manifests itself in various systems, e.g., the centrifugal regulator of James Watt in a steam engine or the thermal (self-organizing) system of atmosphere.

In the Earth's atmospheric thermal system, the input is the temperature T_{gb}, which is determined by solar radiation (for the distance of Earth from the Sun) whereas the output is the average surface temperature, T_s; for the Earth, the temperature $T_{gb} = 288.2$ K or 15 °C. Thus, the average surface temperature, T_s; is a linear function of T_{gb}, which characterizes the solar radiation at the distance from Earth to Sun. This enables one to determine the average temperature, T, at any point in the troposphere with pressure, p:

$$T = T_{gb}\left(\frac{p}{p_o}\right)^{\alpha}. \qquad \text{(Eq. 9.5)}$$

In Eq. 9.4, C and α are defined for the temperature $T_{gb} = 288.2$ K and the atmospheric pressure at sea level, one can rewrite Eq. 9.5 in the following form:

$$T_{gb} = Cp_o^{\alpha}. \qquad \text{(Eq. 9.6)}$$

In Eq. 9.6, the constant C is equal to 288.2. One can then define the average temperature at any elevation in the troposphere by the following exponential function at $p > 0.2$ atm:

$$T = 288.2\left(\frac{p}{p_o}\right)^{\alpha}. \qquad \text{(Eq. 9.7)}$$

It is noteworthy that in the formula,

$$T_{bg} = \left(\frac{S}{4\sigma}\right)^{\frac{1}{4}} \qquad \text{(Eq. 9.8)}$$

for computing the temperature of an *absolutely black-body*, the solar constant S is divided by 4, because the area of Earth's disk insolation is four times lower than the total illuminated area of the Earth. Eq. 9.8 is valid only if the axis of rotation of the planet is strictly perpendicular to the ecliptic plane and the angle of precession, ψ, is equal to zero.

Effect of the Precession Angle

In our solar system, the angle of inclination of the equatorial plane to the ecliptic plane is not equal to zero for either the Sun or the Earth and it changes seasonally with time. Therefore, each of the Earth's polar regions is insolated for half a year. The other half of a year, it is deprived of the influx of solar energy. When one of the polar regions is insolated, the other lies in the shadow of the Earth's body and cannot receive solar energy. The remainder of the Earth's surface receives its portion of solar energy on a regular basis and, consequently, Eq. 9.8 is valid for the calculation of temperature. Therefore, in computing the average temperature of an *inclined* planet at high latitudes (polar regions) one needs to divide the solar constant by 2 (not by 4). In addition, one must take into consideration the spherical shape of polar region. As a result, the solar constant, S, in Eq. 9.8 must be divided by a number, N, which lies between 2 and 4. Considering all the above and that the precession angle is relatively small, one can develop the following equation (see Chapter 2 for derivation)) for the distribution of average temperature in the troposphere,

$$T = \left[\frac{S}{\dfrac{4\sigma\left(\left(\dfrac{\pi}{2}\right)-\psi\right)}{\left(\dfrac{\pi}{2}\right)} + 2\psi\left(\dfrac{\pi}{2}\right)\left(\dfrac{2}{(1+\cos\psi)}\right)} \right]^{\frac{1}{4}} \left(\frac{p}{p_o}\right)^{\alpha} \qquad \text{(Eq. 2.25)}$$

where ψ is the Earth's precession angle; p is the atmospheric pressure at a given altitude, e.g., 0.2 atm $< p < P_0$), $P_0 = 1$ atm, and $\alpha = (c_p / c_o)$ where c_p and c_0 are the specific heats of atmosphere at constant pressure and constant volume, respectively

Today, the Earth's precession angle is $\psi = 23.44°$ *and* the Earth's average surface temperature is $T_s \approx 288.2$ K at sea level. This implies, that the average surface temperature of Earth, considering the Earth's precession, is equal to the efficient temperature of an effectively *gray-body* at the average distance of the Earth to the Sun, ($T_s = T_{gb}$). Thus, the average temperature of contemporary troposphere at any altitude, where $p > 0.2$ atm, is equal to:

$$T = T_{gb}\left(\frac{p}{p_o}\right)^{\alpha} = 288.2\left(\frac{p}{p_o}\right)^{\alpha} \quad \text{(Eq. 9.9)}$$

If the axis of the Earth's rotation were perpendicular to the ecliptic plane, then, at $p = 1$ atm, the average surface temperature would be equal to 278.6 °K, the temperature of *absolutely black-body* at the distance from Earth to Sun. This difference in temperature for $y \approx 23.44°$ and $\psi = 0°$ approaches 9.6 to 10 °C. Consequently, the present-day radiation temperature, considering the precession angle, for the distance from the Earth to the Sun is equal to 263.6 K and not 255 K. Under the classical concept, the precession angle, $\Psi = 0$. The Earth's effective temperature, T_e, is 255 °K or −18 °C. The greenhouse effect for the Earth is currently accepted at +33 °C and a precession angle value of $\Psi = 23.44°$. Correcting the precession angle to $\psi = 0$, the effective temperature would be $T_e = 263.5$ °K or −9.5 °C. This gives a corrected greenhouse effect for the Earth that is much lower, $\Delta T = 24$ °C.

The greenhouse effect is a real phenomenon, although the term itself is unfortunate, misunderstood, and often physically wrong. Common belief is that the Earth's atmosphere, containing the so-called greenhouse gases, weakly absorbs energy from the short-wave solar radiation from the Sun, of which ≈51% (see Figure 9.2) reaches the Earth's surface, but impedes the long-wave (thermal) radiation that is reflected from the Earth's surface, thereby significantly decreasing the Earth's energy loss into the outer space. This has been often accepted as the main cause of atmospheric temperature increase. Thus, it is often felt, the higher the concentration of greenhouse gases in the atmosphere, those gases that absorb the thermal radiation, the greater the heating of the atmosphere. The effect of atmospheric heating under the influence of absorption of energy by the greenhouse gases of heat radiation coming from the Earth's surface was called the greenhouse effect.

By analogy for greenhouses covered with glass, the glass allows the visible light of the electromagnetic spectrum to pass through the glass carrying energy; however, when some of the energy of the light ray is absorbed by the surface that it falls on, the ray is then emitted (radiated) as IR radiation which now has a longer wave length (due to loss of energy) and can no longer pass back through the glass, entrapping the energy within the greenhouse. However, the consequence of the greenhouse effect in the Earth's atmosphere is different. The difference is that in the case of the greenhouse, the air is isolated from the air outside the greenhouse which prevents a convective mixing with the outside air. As a result, as soon as greenhouse windows are opened and the connection to the air outside space is restored, air convection immediately cools the inside of the greenhouse and any greenhouse effect vanishes from the greenhouse.

Convective Heat Transfer in Troposphere

The domination of the convective component in the heat loss by the troposphere has a natural explanation. Upon absorption of energy by the greenhouse gases, the light wave itself vanishes as its energy passes to the oscillatory motions of gas molecules, i.e., to heating of the irradiated gas volume. The heated gas expands and becomes less dense, the air mass then rises toward the stratospheric strata (by buoyancy) where the energy (heat) is dispersed through convection. Upon cooling, this gas in the atmosphere then descends, restoring the air to the previous or even a lower temperature. An analogous situation occurs with heating air containing moisture which condenses within it. The rate of convection (removal of heat from air) is always many orders of magnitude greater than the diffusion rate of heat transfer.

Effect of Water Vapor on Heat Transfer

Water vapor is a greenhouse gas that plays an important part in the temperature distribution within a cloudy troposphere. The temperature distribution, below the cloud cover (see Figure 9.2) may be defined by the humid-air adiabatic curve, and above it, by a dry-air adiabatic curve. The temperature distribution within the cloud cover depends upon the energy released during the moisture condensation process. The average mass of water vapor in the atmosphere, m_u, may be determined from the average atmospheric moisture heat absorbing capacity of the condensation process, $C_w = 0.0791$ cal/g · K. The heat absorbing capacity of 1 gram of water vapor is equal to 0.47 cal/K. The moisture content above 1 cm^2 area of the Earth's surface:

$$m_w \approx \frac{0.49}{0.0791} \approx 6.2 \text{ g/cm}^2. \qquad \text{(Eq. 9.10)}$$

The moisture content, (H_2O), within the atmosphere, near the Earth's surface is ≈0.62%. The heat absorbing capacity of humid and heat-absorbing air is:

$$\sum C_c = (c_p + C_w + C_r) = 0.2394 + 0.0791 + 0.0412$$
$$= 0.3597 \text{ cal/g} \cdot \text{K}, \qquad \text{(Eq. 9.11)}$$

whereas total heat absorbing capacity of dry and IR radiation absorbing air is:

$$\sum C_r = (c_p + C)_r = 0.2394 + 0.0412 = 0.2806 \text{ cal/g} \cdot \text{K}. \quad \text{(Eq. 9.12)}$$

If the cloud cover extends from one to 2.5 km altitude, then the specific air mass within this cloud cover is approximately equal to the difference of pressures at its boundaries:

$$m_{cl} = \Delta p = 887 - 737.1 = 149.9 \text{ g/cm}^2, \quad \text{(Eq. 9.13)}$$

and the moisture mass within the cloud stratum is equal,

$$m_{w_{cl}} = 149.9\left(\frac{0.0062}{2}\right) \approx 0.465 \text{ g/cm}^2. \quad \text{(Eq. 9.14)}$$

The moisture condensation internal heat, $q = 595.8$ cal/g, the average air heating in the cloud stratum is:

$$\Delta T = \frac{q m_{w_{cl}}}{m_{cl}(C_w)_s} \approx 5.14°\text{C}. \quad \text{(Eq. 9.15)}$$

The temperature distribution within the troposphere with a cloud cover is illustrated by Figure 9.2. In the case of cloud cover, it is likely air temperature inversions may emerge even without a horizontal transfer of air mass.

Effect of Carbon Dioxide on Temperature Distribution

Looking at the carbon dioxide effect on temperature distribution in the planetary troposphere, the air pressure distribution in the troposphere is defined by the following exponential equation:

$$p = p_o e^{-h\frac{g\mu}{RT}} \quad \text{(Eq. 9.16)}$$

Then, at $p_0 = 1$ atm:

$$h = \frac{-RT}{g\mu} \ln p \quad \text{(Eq. 9.17)}$$

where R = 1.987 cal/deg · mol is the gas constant; g = 981 cm/s^2 is the gravity acceleration; μ is molecular weight of the air gas mixture in degrees Kelvin, *K*, at the elevation *h*.

Eq. 9.7 determines the temperature distribution in the Earth's troposphere for air mixtures of various molar weights. Figure 9.8 is a comparison of the temperature distribution for a normal Earth type nitrogen-oxygen atmosphere with (1) the molar weight, $\mu = 29.89$; (2) an adiabatic exponent, $\alpha = 0.1905$; (3) a hypothetical carbon dioxide atmosphere with the molar weight $\mu = 44$; (5) a pressure, $p = 1$ atm; and (6) a α exponent determined from Eqs. 2.14 and 2.15. At $\alpha = 0.1428$, the, coefficient, $b^{\alpha} = 1.597^{0.1428} = 1.069$, and the near-surface temperature of this hypothetical carbon dioxide atmosphere is 281.6 K, which is 6.4 °C lower than for the nitrogen-oxygen atmosphere.

Figure 4.6 shows that the temperature distribution curve for a carbon dioxide atmosphere is always below that of the nitrogen-oxygen atmosphere curve. The near-Earth surface temperature for the carbon-dioxide atmosphere *curve 2* turns is lower than the nitrogen/oxygen *curve 1* by ≈6.4 °C than the nitrogen-oxygen atmosphere and not substantially higher as it is assumed by orthodox ecologists. Thus, substantial amounts of carbon dioxide in atmosphere will only cause cooling of the Earth, whereas relatively minuscule changes in the CO_2 partial pressure (by hundreds ppm) have had practically no effect on the temperature of the tropospheric and Earth's surface as seen in measurements over the past 20 years.

Similarly hypothesizing the imaginary replacement of the carbon dioxide Venus atmosphere with a nitrogen-oxygen one at the same pressure of 90.9 atm, the surface temperature would rise from 735 to 796 K (462 to 523 °C). This demonstrates that the atmosphere saturation with carbon dioxide, with all other conditions equal (atmospheric pressure), would always result in a decrease of the greenhouse effect and average temperature within the troposphere. The reason for this is simple; the carbon dioxide molar weight is 1.5 times greater and its heat-absorbing capacity is about 1.2 times lower than for Earth's atmospheric air. As a result, as it follows from Eq. 2.15, the adiabatic exponent for the Venus carbon dioxide atmosphere, with all other conditions equal, is lower, $\alpha_{(CO2)} = 0.1786$ (approximately by a factor 1.067), than for the nitrogen-oxygen composition air, $\alpha_{(N2+O2)} = 0.1905$. The increase in the heat absorption by carbon dioxide will result in an increase of the correction factor C_r and resulting in an additional decline of the adiabatic exponent, $\alpha_{(CO2)}$. This, will cause an additional temperature decline. As noted in the imaginary experiment with Earth's atmosphere, equal pressures as defined by Eq. 9.7, for a carbon dioxide and nitrogen-oxygen atmospheres, are positioned at different elevations. Because a carbon dioxide atmosphere's molar weight is heavier, its equal pressure level is always lower than that for a nitrogen-oxygen atmosphere. This results in a higher temperature distribution for a Venus nitrogen-oxygen troposphere.

The Effect of Carbon Dioxide Anthropogenic Emissions

Various estimates of CO_2 anthropogenic emissions resulting from burning hydrocarbon fuels (2005 data) produce 5 to 7 billion tons of carbon dioxide or 1.4 to 1.9 billion tons of pure carbon to the atmosphere. It has been thought by some that the amount of carbon entering the atmosphere in recent years may have even reached 20 to 35 billion tons. Some people believe that this colossal amount of carbon entering the atmosphere can not only alter the composition of its gas mixture and decline in its heat absorbing capacity, but also increase the total atmospheric pressure.

These two factors operate in opposite directions. As a result, in theory, the average Earth's surface temperature will almost remain unchanged. The Earth's temperature will not change, even if carbon dioxide concentration in the atmosphere doubles, which is currently anticipated by some in 2100. One should also consider that the major portion of carbon dioxide entering the atmosphere, under Henry's law, dissolves in the oceanic water and in hydration of Earth's crust carbon dioxide is bonded in carbonates. Thus, along with carbon, part of the atmospheric oxygen passes into carbonate rocks. Therefore, rather than a slight increase in the atmospheric pressure, we are likely to see a slight decrease, and therefore a slight climate cooling.

The carbon dioxide partial pressure in the atmosphere is regulated by the ocean water temperature. Carbon dioxide solubility in water is controlled by Henry's law:

$$C_{(CO_2)} = p_{CO_2}\left(H \cdot e^{-\frac{-\Delta H}{RT}}\right), \qquad \text{(Eq. 9.18)}$$

where p_{CO2} is the atmospheric carbon dioxide partial pressure; $C_{(CO2)} = 0.1544 \cdot 10.^3$ is the carbon dioxide concentration in oceanic water (HCO_3^{1-} plus CO_3^{2-}) (Kokin *et al.*, 1990); H is the Henry coefficient; ΔH is the enthalpy of carbon dioxide dissolution in the water process; $R = 1.987$ cal/mole $\cdot$ K is the gas constant; and T is upper oceanic water-layer temperature, deg. K. Eq. 9.18 states that the CO_2 partial pressure in the atmosphere will increase when water is heating and decrease when cooling (this is well known to many as the Champaign effect). Figure 9.4 demonstrates the solubility of carbon dioxide gas in water.

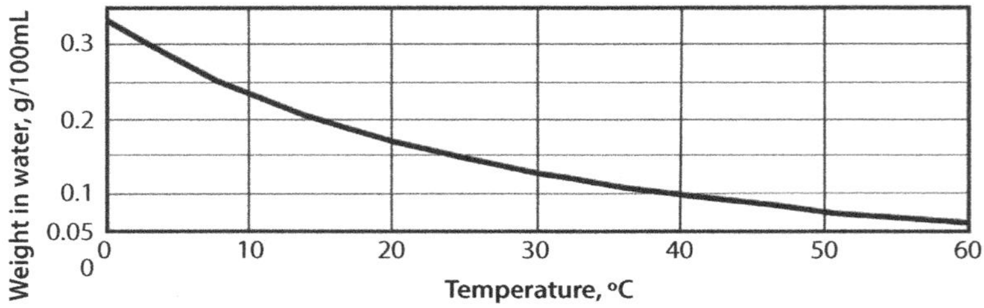

Figure 9.4 Relationship between the temperature and the solubility of CO_2 dissolved in water. (After Wallace, 2009.)

The enthalpy of carbon dioxide's dissolution in water is controlled by the difference between the enthalpy of carbon dioxide solution in water and the enthalpy of CO_2 gas phase: $\Delta H = -98.9 - (-94.05) = 4.85$ kcal/mole (Naumov *et al.*, 1971). Eq.9.8 may be solved for an earlier time period by assuming a preindustrial value for the CO_2 partial pressure and temperature, e.g., a preindustrial time like 1880: $p_{(CO2)} = 2.9 \times 10^{-4}$ bar and $T = 287.2$ K. In this case $H = 10^{-4}$ 1/bar. Currently carbon dioxide amount dissolved in oceanic water is approximately 60 to 90 times its amount in the atmosphere. For calculations it may be assumed $C_{(CO2)} \approx$ const. Even a slight increase in the average temperature of oceanic water results in a notable increase in the atmospheric carbon dioxide partial pressure. A significant increase in $p_{(CO2)}$ was observed due to the anthropogenic influence (on the order of $\Delta p = 60$ ppm). A more modest contribution was introduced by the increase in Earth's surface temperature from 287.2 K to 288 K, $Dp = 6.9$ ppm. A natural warming during the recent 120 years by 0.8 °C would result in an increase in carbon dioxide partial pressure by $\Delta p = 6.9$ ppm. Anthropogenic emissions added 53 ppm more, although this addition likely had nothing to do with the observed climate warming as this warming was associated with a change in the Sun's activity. It has been noted that atmospheric carbon dioxide concentration fluctuations in the Pleistocene occurred at changes of glaciation epochs, by glaciation interstadials, and they exceeded 150 ppm, naturally, without any anthropogenic intervention.

Similar conclusions to those above have been arrived at by many scientists studying climate change. Robinson *et al.* (1998) and others have reported no climate warming at all. Seitz has written: "Experimental data of climate change do not show any harmful effect from the anthropogenic use of hydrocarbons. Contrary to that, there is evidence that increases in carbon dioxide content in atmosphere may be beneficial". F. Seitz prepared a petition by scientists to the U.S. government with an appeal to reject the international agreement on the global climate warming concluded in Kyoto (Japan) in December of 1997 and other similar agreements. The petition stated: "There are no convincing scientific evidences that the anthropogenic emissions of carbon dioxide, methane and other greenhouse gases are incurring or may in the foreseeable future cause a catastrophic warming of Earth atmosphere and destruction of its climate. Besides, substantial scientific evidences are available, which indicate that an increase of carbon dioxide concentration in the atmosphere renders positive effect on the natural growth of plants and animals in Earth's ambient medium". By 2005, almost 17,000 U.S. scientists and engineers had signed this petition.

The conventional concept of climate warming due to the accumulation, in the troposphere, of anthropogenic CO_2 and other greenhouse gases is a myth and has no scientific support. The accumulation of CO_2 in the atmosphere has historically shown no noticeable effect on the temperature regime of Earth's climate. Temperature changes are the result of primary causes, whereas carbon dioxide concentration variations (except for the anthropogenic influence) are secondary causes and are the consequence and not the cause of temperature changes. Peaks in the Sun's irradiation preceed the peaks in CO_2 concentration in the atmosphere.

Greenhouse gas, especially carbon dioxide and water vapor, accumulation in the troposphere can increase somewhat the convective air mass heat exchange in the troposphere. This, in turn, may cause intensification of the synoptic processes and storm activity within the troposphere. Strengthening of hurricanes and tornados, in recent years, incurring substantial damages in the southern portion of the United States, should only be associated with fluctuations in the solar activity. Maximum solar activity (sunspots) were observed in 2000–2002; however, due to a high heat absorbing capacity of water in the surface oceanic stratum, the climatic reaction to a change in solar activity was delayed by a few years, relative to its 11-year cycle. That may be the reason why maximum activity of the synoptic processes moved to the initial years of the twenty-first century.

10

Development of Carbon Dioxide in Earth's Atmosphere

Carbon Dioxide

Carbon is the fourth most abundant element in the universe and is a major component for life on the Earth. On the Earth, organisms utilize carbon for structure and/or energy. For a molecule to be organic, it must contain carbon atoms. Carbon is found in a diverse variety of forms: (1) as a gas, e.g., carbon dioxide (CO_2); (2) a liquid, e.g., petroleum hydrocarbons; and (3) as a solid, e.g., carbonates ($CaCO_3$), wood, plastics, diamonds, coal, peat and graphite.

Carbon dioxide is a colorless, odorless gas that is found (1) in the atmosphere as a trace gas, (2) as a gas dissolved in water (oceans), (3) in liquid hydrocarbon deposits as a dissolved gas and (4) in volcanic activity (CCS, 2017).

Carbon dioxide has a specific gravity of 1.555 or is ≈60% heavier than air. The carbon dioxide molecule consists of a carbon atom covalently bonded to two oxygen atoms. It occurs naturally in the atmosphere, and is soluble in water, ethanol and acetone. It is a linear covalent molecule. In the atmosphere, it reacts readily with water droplets in clouds to form carbonic acid, which has a pH of 6.8:

$$CO_2 + H_2O \rightleftharpoons H_2CO_3 \qquad \text{(Eq. 10.1)}$$

It is a very stable molecule; few chemical processes other than photosynthesis have been found that are able to effectively reduce carbon dioxide to carbon monoxide (CCS, 2017).

Although carbon dioxide is viewed by some as a poisonous gas, life on Earth is dependent upon it. It is required for those organisms that use it in photosynthesis to harvest energy and produce carbohydrates and oxygen.

Sources of Carbon Dioxide

Archer (2010) noted that there are several different forms of carbon that one can track in the global carbon cycle:

1. Inorganic-carbon (C) in rocks, e.g., bicarbonate, carbonate, and other rocks.
2. Organic-carbon, e.g., organic plant material, hydrocarbons, etc.
3. Carbon gases (CO_2), e.g., volcanism, degassing of Earth's mantle material.
4. Solution gas, (CO_2), e.g., dissolved in the ocean water.
5. Methane gas (CH_4) of non-organic origin.
6. Carbon monoxide (CO), e.g., found in atmosphere.

The Carbon Cycle

Today's movement of carbon through the Earth's atmosphere, hydrosphere, biosphere and geosphere (which is only a portion of the total *Carbon Cycle*) is shown in Figure 10.1. This portion of the carbon cycle displays several reservoirs of carbon storage and the various processes and forms by which the carbon moves between these reservoirs. The major carbon reservoirs displayed in this figure include the atmosphere, oceans, vegetation, and soil. Also shown is the approximate quantities of carbon that flow in and out of the specific storage areas per year. The arrows indicate the direction of the carbon flow between these carbon storage reservoirs. Carbon dioxide is the primary carrier for the movement of carbon between the storage

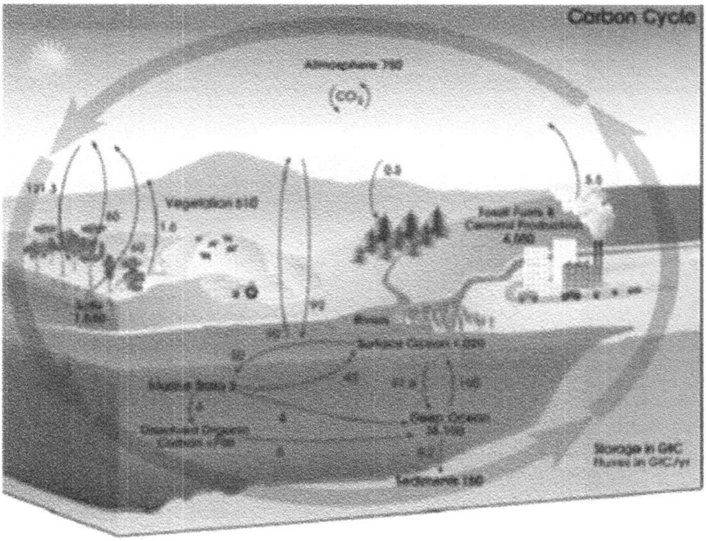

Figure 10.1 A portion of today's carbon cycle showing the carbon path through the atmosphere, oceans, and soils. (After Harrison, 2003.)

reservoirs. There are several smaller cycles or pathways of carbon flow as shown in Figure 10.1.

The ocean/atmospheric cycle is a major pathway between the two large storage reservoirs of carbon dioxide. When the water temperature increases in the oceans (which cover 72% of the surface of the Earth), the solubility of the carbon dioxide gas in the water (dissolved gas) deceases and large volumes of carbon dioxide, previously dissolved in the ocean water, are released to the atmosphere.

The *carbon rock cycle* is a description of another portion of the carbon cycle, including the movement of carbon through (1) degassing of the mantle (2) plate tectonics, (volcanism), and (3) carbonate rock formation. This is often referred to as the *Rock Cycle* (Figure 10.2). Carbon dioxide is the primary form of carbon circulating through the Earth's atmosphere, coming from outgassing of the Earth's mantle at the midocean ridges, hotspot volcanoes and subduction-related zones. Archer (2010) noted that over geologic history, there is a close relationship of the percent content of carbon dioxide in the atmospheric and volcanic activity (Figure 10.3). During high periods of volcanic activity, the percent atmospheric carbon dioxide concentration is higher; therefore, the increased percentage of carbon dioxide in the atmosphere can be partially related to volcanic activity. Volcanic activity from 1850 to 2010 is shown in Figure 10.4. Much of

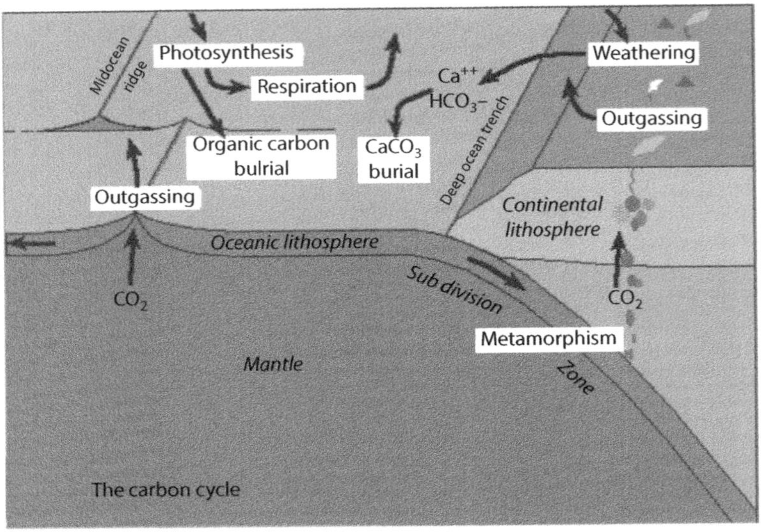

Figure 10.2 The rock cycle portion of the carbon cycle. The primary source of CO_2 is degassing from the Earth's interior. (After Columbia, 2017, http://www.columbia.edu/~vjd1/carbon.htm.)

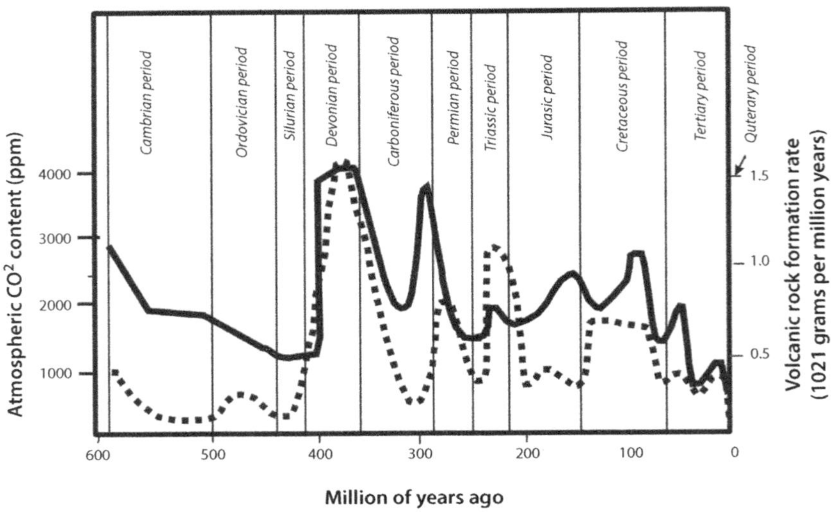

Figure 10.3 Schematic showing relationship between the CO_2 content in the atmosphere (dashed line) and the volcanic activity on Earth (solid line). (Modified after Archer, 2010.)

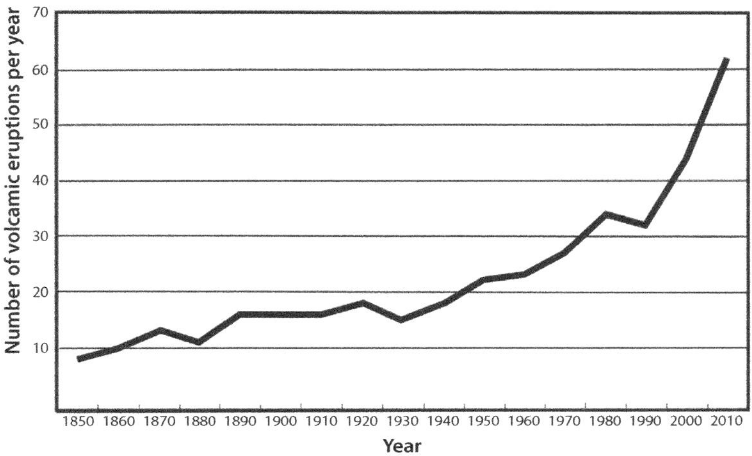

Figure 10.4 Global volcanism over time (1850 to 2010). There has been a general increase in volcanism over the past 100 years. (Modified after http://informacaoincorrecta.blogspot.com/2011/08/porque-nibiru-nao-existe.html.)

this carbon is abiogenic carbon (inorganic gas), derived from (1) the subduction and metamorphism of carbonate rocks and (2) degassing of the mantle. The percentage of carbon dioxide shown in Figure 5.1 in the early periods of the Earth's history, prior to bacteria, is of abiogenic origin.

Mass of Carbon in the Earth's Crust

Ronov and Yaroshevsky (1978) estimated that $\approx 3.91 \times 10^{23}$ g of CO_2 is bonded in the crustal carbonates. Cores also contain $\approx 1.95 \times 10^{23}$ g of organic carbon, C_{org}. The organic carbon, C_{org}, prior to its reduction in the biologic processes, was bonded by 5.2×10^{22} g of oxygen. Therefore, the total CO_2 mass buried in the crust is:

$$m(CO_2) \approx (3.91 + 0.72) \times 10^{23} = 4.63 \times 10^{23}\, g.$$

Mass of Carbon in the Earth's Mantle

The determination of carbon content in the mantle is much more complicated than for the Earth's crust. Experimental data indicates that the high-temperature distillation fractions of volatile components in an oxygen atmosphere of tempering glass from the tholeiitic basalts in the oceanic rift zone contain 20 to 170 g/ton of mantle-derived carbon having isotopic shifts of about −5‰ (Watanabe *et al.*, 1983; Sakai *et al.*, 1984; Exley *et al.*, 1986). Dispersed carbon is found in all igneous rocks in negligible concentrations of 10 to100 g/ton with a substantial deficit of the heavy isotope $\delta^{13}C \approx 22$ to 27‰ (Galimov, 1988a). Carbon found in the Earth's crust is isotopically heavier, with $\delta^{13}C \approx -3$ to −8‰ (average value is −5‰). Part of the mantle carbon is dispersed among the silicate crystalline grids in the atomic state (Watanabe *et al.*, 1983) and, hence, is not a volatile component of the basalt melts. Therefore, carbon content in the carbon dioxide gas phase of basalt melt-outs may be noticeably smaller than the quoted total carbon content.

The assumption here is that the mobile carbon content in the mantle is ≈30 g/ton or 110 g/ton when normalized for carbon dioxide. In this situation, the mantle carbon dioxide content is $m_{CO2} \approx 4.48 \times 10^{23}$ g of carbon dioxide. Carbon dioxide degassing of the mantle, as well as degassing of the other mobile volatile components of the atmosphere and hydrosphere, may be described by the relationship:

$$m_i = m_{i_o}\left(1 - e^{-Xz}\right) \qquad \text{(Eq. 10.2)}$$

where m_i is the mass of the degassed volatile component *i*; m_{i_o} is the total mass of the component *i* on he Earth; × is the component *i*'s mobility exponent; *z* is the tectonic parameter defined by the ratio of total loss of the Earth's depth heat and its current value $Q_0 = 13.42 \times 10^{37}$ erg (Sorokhtin and Ushakov, 2002). Today's value of *z* is equal to 1 (see Figure 10.5).

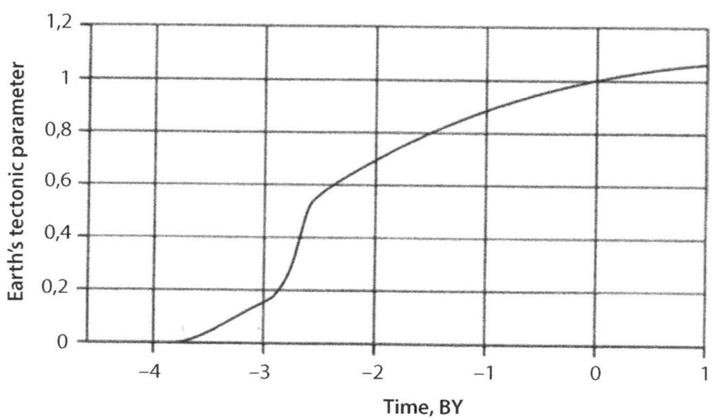

Figure 10.5 Earth's tectonic parameter defining the mantle degassing rate.

For solving Eq. 10.2, one needs to insert the marginal conditions. Assuming, as one of the marginal conditions, that ≈3.91 × 10^{23} g of carbon dioxide are bonded within the carbonates of the Earth's crust with additional 1.95 × 10^{22} g of organic carbon, C_{org} (Ronov and Yaroshevsky, 1978). Prior to the reduction of the organic carbon (C_{org}) by biological processes, 5.2 × 10^{22} g of oxygen was bonded along with it. Therefore, the total mass of carbon dioxide degassed from the mantle is:

$$m_{CO_2} \approx (3.91 + 0.72) \times 10^{23} \approx 4.63 \times 10^{23}\ g.$$

It is harder to determine the carbon dioxide content of the Earth's mantle.

The experimental data indicates the following: (1) High-temperature distillation fractions (distillation in an oxygen atmosphere) from tempering glasses of the oceanic rift zones' tholeiite basalts usually include 20 to 170 g/ton of carbon of mantle origin carbon with an isotope shift of about −5‰ (Sakai *et al.*, 1984; Watanabe *et al.*, 1983). (2) The scatter of experimental data is too substantial for a reliable determination of average carbon and CO_2 content in the mantle. (3) The mantle carbon is in an atomic state, dispersed in silicates' crystalline grids (Watanabe *et al.*, 1983) and, therefore, not a volatile component of basalt melts. Therefore, carbon content in the CO_2 gas phase of basalt melt may be notably lower than quoted here. One can assume that the mobile carbon content in the mantle is around 30 g/ton or, recalculated for carbon dioxide, 110 g/ton. In this case the mantle contains nearly $m_{CO_2} \approx 9.11 \times 10^{23}\ g$. The following relationship can be used:

$$C_{CO_2} = p_{CO_2} \cdot He^{\frac{-\Delta H}{RT}} \qquad \text{(Eq. 10.3)}$$

where p_{CO_2} is the carbon dioxide partial pressure in the atmosphere; C_{CO_2} $(\approx 0.1544 \times 10^{-3})$ is the carbon dioxide concentration in the ocean water (Kokin *et al.*, 1990) $(HCO_3^- + CO_3^{2-})$; H is the Henry's coefficient; ΔH is the entropy of the carbon dioxide dissolution in water; and T is the water temperature in deg. K.

Another aspect of Eq. 10.2 is that the Earth's degassing rate is

$$\dot{m} = m_m \chi \dot{z}. \qquad \text{(Eq. 10.4)}$$

where the derivative of the tectonic parameter $\dot{z}$ defines the Earth's tectonic activity (see Figure 10.5). For this situation, $\chi(CO_2) \approx 0.71$.

The total mass of carbon dioxide in the Earth, with C_{org} normalized for CO_2, is $m_{CO_2} \approx 9.11 \times 10^{23}$ g.

Using Eq. 10.4, one can plot the curve of carbon dioxide degassing from the mantle as shown in Figure 10.6, which indicates that the carbon dioxide degassing rate peak coincides with the period of the Earth's maximum tectonic activity about 2.7 BY ago (Figure 10.7).

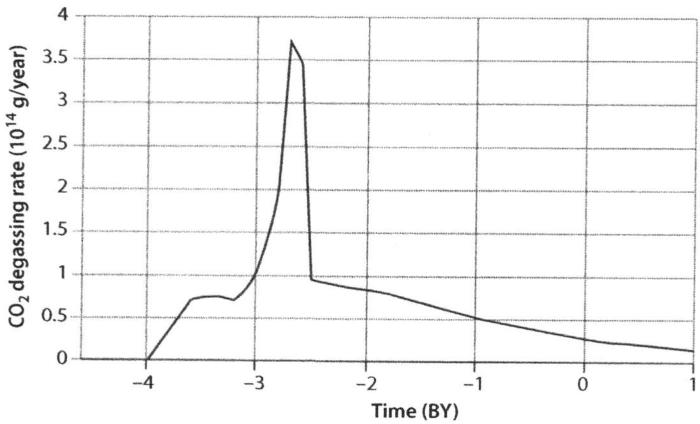

Figure 10.6 Rate of carbon dioxide degassing from mantle with time.

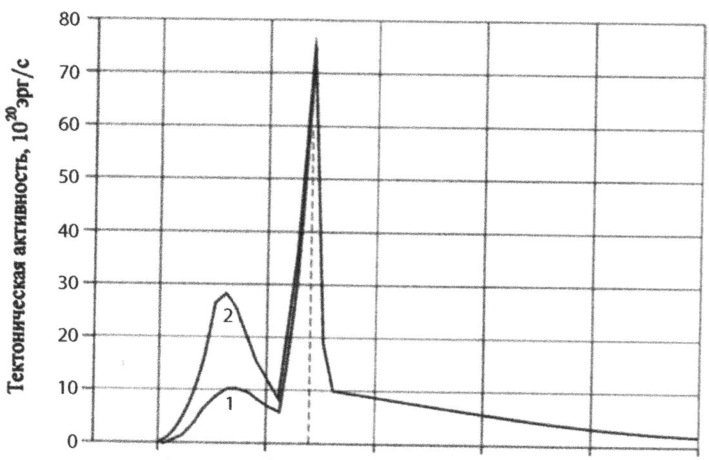

Figure 10.7 Earth's tectonic activity measured by the heat flow from the mantle, m().zQ∝: *Curve1* – Average for the Earth as a whole. *Curve 2* – Tectonic activity within a wide ring belt of the Archaean Earth crust formation above a differentiation zone of the Earth's matter. Dashed vertical line shows the time of the Earth's core separation.

Historic Content of Carbon Dioxide in the Earth's Atmosphere

The percentage of carbon dioxide in the Earth's atmosphere has changed over the geologic history of the Earth (see Figures 5.1 and 10.3). One reason the total atmospheric pressure of the Earth has varied is due to the changing composition of its various gases making up the atmosphere. When the Sun, Earth and other planets were formed, by coalescing of matter from a rotating nebula (≈4.6 BY ago), during the period of the accretion of particles forming the Earth, many volatile and nonvolatile elements from the nebular were present. Due to the centrifugal forces, the accumulation of mass at the center of this rotating planet gathered materials of heavier mass, e.g., iron, etc., whereas the lighter-gaseous elements formed the outermost layer of the Earth (atmosphere). These lighter-gaseous elements were composed of hydrogen, helium, nitrogen and other volatile-light gases of the condensing planetary nebula material.

Earth's Hadean Atmosphere (4.56 to 4.0 BY ago)

At the beginning of the Hadean time, the Earth's surface consisted of molten rock, a magma ocean and very little atmosphere. Due to the Earth's high surface temperature, water existed only in the atmosphere as vapor. Within this hot atmosphere, the lightest elements, the hydrogen and helium molecules had an average speed greater than the velocity required to escape from the Earth's gravity. Thus, they escaped from the atmosphere into space. As a result, the Earth's Hadean atmosphere lost its hydrogen and helium and all that was left in the atmosphere were the molecules of methane, ammonia, water vapor, and small percentages of nitrogen and carbon dioxide. A cataclysmic meteorite bombardment (≈3.9 BY ago) kept much of the Earth's surface in a molten state with elevated surface temperatures. The incoming impactors likely brought with them additional water, methane, ammonia, hydrogen sulfide and other gases that supplemented the Earth's earliest atmosphere as indicated in Figure 5.1.

The high surface temperature of the Earth during the Hadean time depleted the atmospheric methane through the endothermic reaction as follows:

$$CH_4 + H_2O \rightarrow CO_2 + 3H_2 \qquad \text{(Eq. 10.5)}$$

This reaction, generating carbon dioxide, requires elevated temperatures of approximately 700 °C to 1100 °C, which would have been common in the hot crust and magma lakes of the Hadean Time. The produced carbon dioxide, readily combined with metals to form carbon compounds.

Volcanic activity began toward the end of the Hadean time, which is reflected by the increasing percentage of carbon dioxide in the atmosphere. As the Earth cooled, the surface changed from molten lava to solid rock and the Earth cooled enough for liquid-water to begin to accumulate at the higher elevations on its surface.

The major carrier of carbon in the *initial stages* of Earth's evolution during the Hadean and Archaean time, was carbon dioxide and the products of its chemical reactions. After the early depletion of methane and ammonia from the Earth's atmosphere, carbon entered the crust and the atmosphere primarily through mantle degassing.

Earth's Archaean Atmosphere (4.0 to 2.4 BY ago)

During the Archaean time, the percentage of carbon dioxide in the atmosphere increased, originating from degassing of the mantle (see Figure 5.1). The volume of volcanic carbon dioxide in the atmosphere peaked and started to decrease as the carbon dioxide was bonding with carbonate minerals. Thus, it is important, that the increase of carbon in mantle mass, M_m, at the beginning of the Archaean ($M_m = 0$) to its amount at the end of the Archean ($M_m\Sigma$) be considered. The increase of the degassing from the mantle mass is estimated in Figures 10.6 and 10.7. The total volatile component of the mantle, m_i, can be estimated as follows:

$$m_i = m_{i_o}(1 - e^{-\chi_i z})\left\{\frac{M_m \cdot t}{M_{m\Sigma}}\right\}, \qquad \text{(Eq. 10.6)}$$

where $M_m \cdot t$ is the mantle mass for the time $4 < t < 2.6$ BY ago. For the Proterozoic and Phanerozoic time, Eq. 10.6 can be used. It is necessary to insert the initial and boundary conditions in Eqs. 10.2 and 10.6 for the content of each component (water, nitrogen or carbon dioxide) in the Earth's external geospheres.

The rate and mass of carbon dioxide from mantle degassing is shown in Figures 10.6 and 10.7. The carbon dioxide accumulation rate in the Earth's geospheres (including the atmosphere, hydrosphere, and crust), is presented in Figure 10.8 (based on Eqs. 10.2 and 10.6).

Mantle carbon dioxide degassing is different from that of the water degassing. The reason is the lower heat of formation for water (the molecular water, $\Delta H_f^O = 57.8$, for CO_2 $\Delta H_f^O = 94.05$, and for FeO $\Delta H_f^O = 63.6$ kcal/ mol)

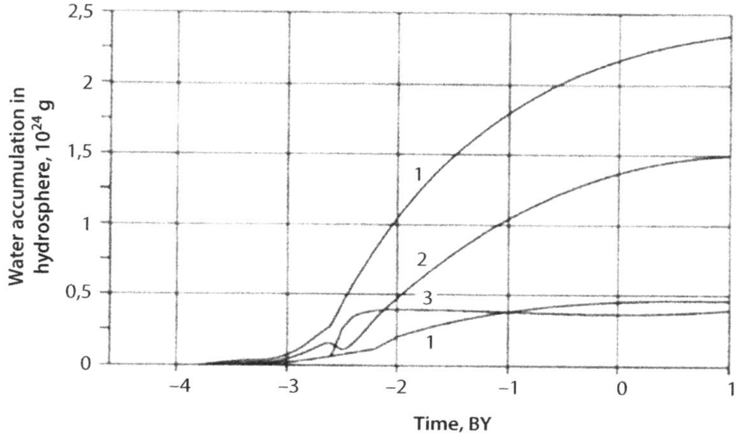

Figure 10.8 Water accumulation in the Earth's hydrosphere: *Curve 1* – Total mass of water degassed from the mantle. *Curve 2* – Water mass in the ocean. *Curve 3* – Water mass bonded in the ocean crust. *Curve 4* – Water mass bonded in the continental crust. (After Sorokhtin and Ushakov, 2002.)

and its dissociation on the metallic iron during the zonal differentiation of the Earth's matter in the Archean time. This CO_2 degassing rate peak coincides with the time of maximum tectonic activity, 2.6 BY ago (see Figure 5.2). At the same time, the water degassing was at its maximum rate ≈2.5 BY ago, that is, after the high-density Earth's core separation and after the switch in the differentiation of the Earth's matter from the zonal melting mechanism to a more quiescent barodiffusion mechanism.

If the entire volume of degassed carbon dioxide had not been bonded, but had been preserved in today's atmosphere as a gas, the Earth's partial pressure would have been ≈91 to 100 atm, which is like today's pressure on Venus, resulting in a condition where life as we know it could not exist. Life on the Earth was fortunate in that simultaneously while CO_2 was entering the atmosphere, it was being bonded into the carbonate rocks. For this reaction to occur, liquid water is required, as only then will silicate hydration be accompanied by carbon dioxide absorption with the emergence of carbonates under the following reactions:

$$4Mg(SiO_4)(\text{forsterite}) + 4H_2O + 2CO_2 \rightarrow \\ Mg(Si_4O_{10})(OH)_8(\text{serpentine}) + 2MgCO_3(\text{magnesite}) \\ + 65.05\ \text{kcal/mol} \qquad \text{(Eq. 10.7)}$$

and

$$2CaAl_2Si_2O_8(\text{anortite}) + 4H_2O + 2CO_2 \rightarrow$$
$$a\ Al_4(Si_4O_{10})(OH)_8(\text{kaolin}) + 2CaCO_3(\text{calcite}) + 10.54\ \text{kcal/mol.} \quad \text{(Eq. 10.8)}$$

Therefore, the evolution of carbon dioxide in the ocean was significant in the development of the carbon dioxide partial pressure in the atmosphere (Figure 10.8).

As shown in Eqs. 10.7 and 10.8, for each two CO_2 molecules bonded in carbonates there are four water molecules used for the hydration of the rock-forming minerals of the oceanic or continental crust. Thus, when carbon dioxide is in excess within the atmosphere and hydrosphere, almost all rock hydration reactions are accompanied by CO_2 bonding in carbonates. Each unit of mass for water bonded in rocks corresponds with approximately 1.22 carbon dioxide mass units. For a carbon dioxide shortage, the hydration of portion of silicates occurs without bonding of carbon dioxide, e.g.,

$$4Mg_2SiO_4(\text{fosterite}) + 6H_2O \rightarrow$$
$$Mg_6[Si_4O_{10}](OH)_8(\text{serpentine}) + 2Mg(OH)_2(\text{brucite})$$
$$+ 33\ \text{cal/mole.} \quad \text{(Eq. 10.9)}$$

The total mass of water bonded in the Earth's crustal rocks is the sum of *Curves 3* and *4* in Figure 10.9. Today, this sum is equal to $(0.358 + 0.446) \times 10^{24} = 0.804 \times 10^{24}$ g. This enables the calculation of carbon dioxide mass, which could have been bonded in carbonates. This mass turns out to be equal to 9.81×10^{23} g, which is a quantity more than twice that contained within carbonates (3.91×10^{23} g). A significant portion of the Earth's crustal rocks were hydrated in Proterozoic and Phanerozoic time due to a substantial carbon dioxide deficit under Eq. 10.9 without forming carbonates.

Life within the ocean has always been limited by the oceans' phosphorus content. Phosphorus solubility in water is relatively low (Schopf, 1982). Therefore, the absolute mass of the organic matter in the ocean is always proportionate with its mass. The organic carbon mass, converted to carbon dioxide, in the Archaean and Proterozoic time, can be estimated considering that it is approximately proportionate to the mass of water in the global ocean (Figure 10.9, *Curve 2*).

Along with the CO_2 that bonded in rocks and the biomass, part of this carbon dioxide, as anions HCO_3^-, was dissolved in oceanic water. One can

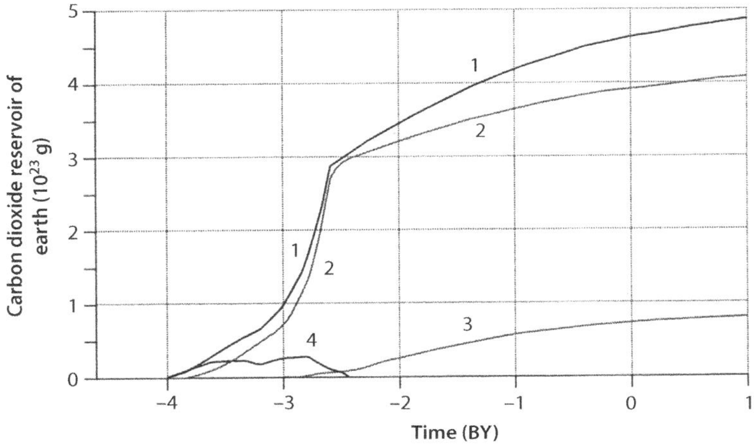

Figure 10.9 Mass of carbon dioxide bonded in the Earth's crust and present in the Archaean atmosphere: *Curve 1* – Mass of CO_2 degassed from the mantle. *Curve 2* – Mass of CO_2 in the Earth's crust carbonates. *Curve 3* – Organic carbon mass normalized for CO_2. and *Curve 4* – CO_2 mass in the Archaean atmosphere.

estimate this portion of CO_2, but it is necessary to remember that carbon dioxide solubility in water, under Henry's law (Sorokhtin and Ushakov, 2002), is directly proportionate with its partial pressure in the atmosphere. The relationship of the carbon dioxide gas solubility versus pressure and temperature in water is presented in Figure 10.10.

In Figure 10.11, the mass of carbon dioxide degassed from the mantle (*Curve 1*) is compared with the mass of carbon dioxide bonded in carbonates (*Curve 2*) and with total carbon dioxide mass in the carbonate and biogenic reservoirs (*Curve 3*). The organic carbon (in conversion to carbon dioxide mass) accumulation in the oceanic deposits is presented by *Curve 5*.

Earth's Proterozoic and Phanerozoic Atmosphere (2.4 BY ago to today)

At the beginning of the Huron Orogeny (≈2.4 BY ago) the carbon dioxide partial pressure in the Early Proterozoic atmosphere drastically declined by about a factor of 10,000 to an equilibrium level of about 1.0 to 1.5 mbar (Chumakov, 1978). The total atmospheric pressure decreased from 5 to 6 bars by the end of the Archean to 1.4 bar in the Early Proterozoic (see Figure 10.12). The entire process of removal of carbon dioxide from the atmosphere during the Archaean/Proterozoic time boundary likely did

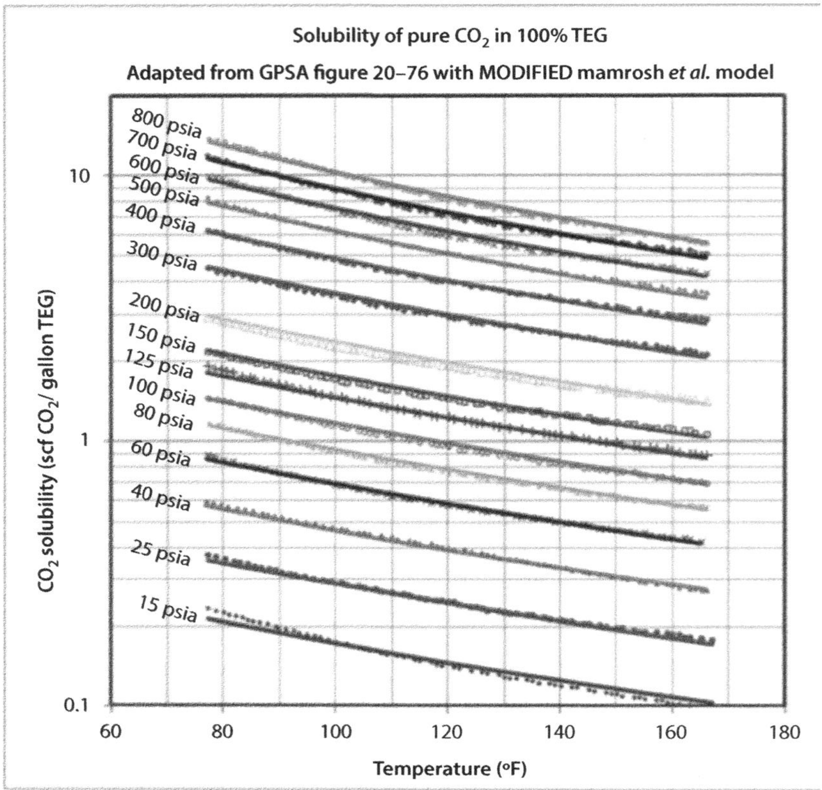

Figure 10.10 Solubility of carbon dioxide in water. (After J. M. Campbell, Petro Skills, 2017, http://www.jmcampbell.com/tip-of-the-month/2012/11/solubility-of-acid-gases-in-teg-solution-part-3-co2-in-teg/.)

not last longer than 100 to 150 MMY. Therefore, the composition of the Early Proterozoic atmosphere became mostly nitrogen with a small content of argon (≈9 mbar). As a result, there was a significant cooling of the Earth, although the average continent stand level remained rather high (4 to 2–3 km). This was a reason for the emergence of the Huron Glaciation: the onset of cooling-down with a high-continent stand (Sorokhtin and Sorokhtin, 1997; Sorokhtin, 2002). Almost all the continents, which at that time formed a unique supercontinent Monogaea, despite their equatorial positions were above the snowline and were covered under a massive ice-sheet glaciation.

Subsequently, the carbon dioxide partial pressure in the atmosphere (see Figure 10.12) was controlled on the average by the oceanic water

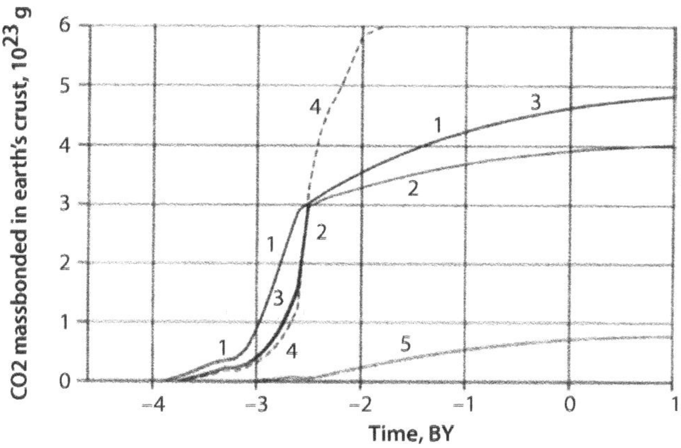

Figure 10.11 Carbon dioxide mass bonded in the Earth's crust: *Curve 1* – Mass of the mantle degassed carbon dioxide. *Curve 2* – Accumulation of carbon dioxide in the Earth's crust carbonates. *Curve 3* – Total mass of bonded carbon dioxide (within the carbonate and biogenic reservoirs). *Curves 1* and *3* superpose and merge. *Curve 4* – Mass of water bonded in Earth's crust. *Curve 5* – Organic carbon mass converted to CO_2.

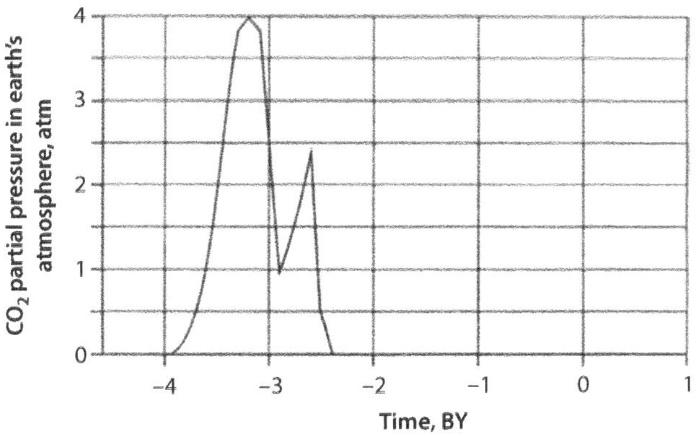

Figure 10.12 Variation in carbon dioxide partial-pressure in the Earth's atmosphere with time.

temperature and Henry's law. During the ice periods, it declined to 0.3 mbar (the current value is approximately 0.46 mbar), and in warm periods increased to 0.7 to 1 mbar.

Figure 10.11 shows that during the Proterozoic and Phanerozoic time, almost the entire carbon dioxide degassed from the mantle was bound in the carbonate or biogenic reservoirs (in the form of C_{org}. The mass of the carbon dioxide in the atmosphere and the hydrosphere of these epochs is impossible to determine. During Archaean, there was a small volume of water in the Archaean oceans, and the mass of bonded carbon dioxide was substantially smaller than the mass of carbon dioxide mass degassed from the mantle. This helps explain the emergence of a carbon dioxide atmosphere during the Archaean.

The separation of Earth's core and the formation of the oceanic crust's serpentinite layer occurred between the end of the Archaean and beginning of the Proterozoic time (≈2.5 BY ago) because of a drastic decline in the Earth's tectonic activity. There was a significant increase in the rate of carbon dioxide bonding in carbonates (Eqs. 10.7 and 10.8), which resulted in the accumulation of oceanic deposits (Sorokhtin and Ushakov, 2002). Therefore, in the Proterozoic and Phanerozoic time, the overwhelming portion of carbon dioxide was bonded in carbonates and organic matter, and only a small portion of carbon dioxide entered the atmosphere and hydrosphere.

As a result, in the Early Proterozoic, the atmospheric carbon dioxide partial pressure sharply declined by a factor of about 1,000 (to a few millibars). This resulted in the accumulation of massive carbonate deposits in the Early Proterozoic. However, as carbonate rocks have low resistance to weathering, most of them were redeposited in younger formations. In the Early Proterozoic there was a significant decline in the carbon dioxide partial pressure in the nitrogen-carbon-dioxide Archaean atmosphere. According to Eq. 2.25, there was a substantial temperature decline and climate cooling of the Earth during the Early Proterozoic. The average temperature of the oceanic, near-bottom water dropped from 70 °C to 30–40 °C (see Figure 1.3). Figure 10.12 displays the carbon dioxide partial pressure evolution in the Earth's atmosphere in consideration of its partial dissolution in oceanic water.

As Figure 10.12 shows, the carbon dioxide partial pressure in the Archaean reached 3 to 4 atm. because there was so little water in the Archaean oceans. The bonded carbon dioxide mass at that time was significantly lower than the mantle degassing of the carbon dioxide mass (see Figure 10.11). Therefore, in the Archaean time, the bulk of the mantle carbon dioxide degassing must have been in the ocean and stored as dissolved gas in the oceanic water. Thus, the oceans were saturated with carbon dioxide. As the average oceanic water temperature was 60–70 °C, the oceanic water was warm, acidic and an aggressive solvent for the ore elements. This resulted in the saturation of the oceanic water by many ore elements, e.g., Au, U, Cu, Pb, Zn and sulphides. When the ocean cooled down and carbon dioxide was bonded with the carbonates, in the Early Proterozoic, the oceanic water remained warm but became chemically neutral (pH ≈ 7). That

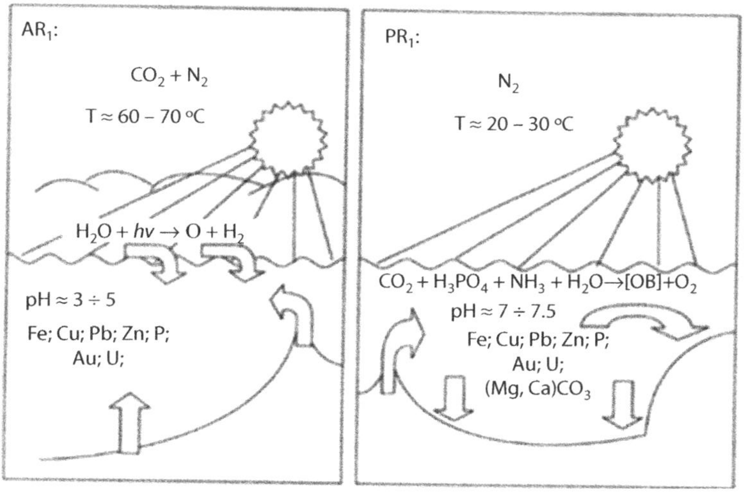

Figure 10.13 The formation of unique stratiform ore deposits in the Early Proterozoic due to the precipitation of ore elements after the bonding of Archaean atmosphere carbon dioxide in carbonates and cooling down of the oceanic water between Archaean and Proterozoic time (OB in the equation is organic matter). The arrows indicate arrival paths of ore elements in the oceans and their precipitation in Early Proterozoic (the descending arrows in Archaean show runoff from continents).

was why the ore elements precipitated (Figure 10.13) forming very rich stratiform deposits of several mineral resources (i.e., a unique gold-uranium deposit, Witwatersrand, South Africa, or copper deposit, Udokan, Siberia).

Anthropogenic Carbon Dioxide in the Atmosphere

The anthropogenic carbon's effect on the Earth's climate has often been evaluated on the assumption that greenhouse gases absorb heat of radiation in the troposphere without the understanding that most heat transfer in the troposphere is by convection. Earlier in this book, it was shown that an increase in the atmosphere's carbon dioxide content does not increase the atmosphere's absorption of energy from the Sun, resulting in an increase in temperature, but rather it decreases the absorption of energy, thus cooling the troposphere.

Various estimates of the volume of anthropogenic carbon dioxide, released by burning of natural fuels are on the order of 7 to 10 billion tons or 1.9 to 2.7 billion tons of carbon per year. This anthropogenic and natural amount of carbon dioxide is shown in the temperature chart (dashed line) for the past 2,000 years (Figure 10.14). An examination of this figure indicates that since the seventeenth century, there may have been a natural increase in temperature of one degree. It also should be noted that the Earth's temperature is cyclic, fluctuating 1 degree over ≈1000-years due to natural, rather than man-made causes. In effect, it is not determinable if there is a temperature increase or decrease over short period of time. Human emissions of CO_2 in Figure 10.14 (dashed lin), show a very steep increase, but at the same time this increase has had a negligible effect on the Earth's rate of temperature increase.

As discussed earlier in this book, a significant increase in the carbon dioxide content in the atmosphere will result in cooling the troposphere, whereas a reduction in the carbon dioxide content would increase the Earth's temperature. Again, this statement is based upon the understanding of how heat is transferred in the troposphere, as most heat in the troposphere is transferred by conduction (adiabatic) and not radiation (see Eq. 2.25 and Chapters 2 and 3). To compute the temperature of a planet's

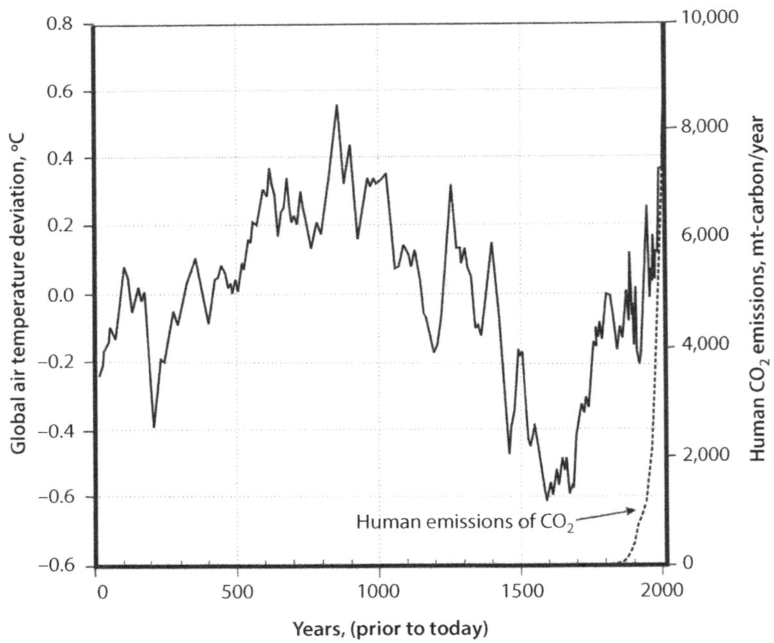

Figure 10.14 Schematic showing relationship between global air temperature and human CO_2 emissions from A.D. 16 to 2010. (Evans, 2010, states that these emission figures are not perfect because they omit some minor causes, e.g., deforestation; however, these are relatively minor). Evans also noted that the temperatures frm 1850 to 1980 are suspect because they were obtained from land-thermometers. (Modified after Evans. 2010.)

atmosphere, consisting of a variety of gases, one only needs to know the components (type gases) and their contents in the atmosphere along with total pressure of the atmosphere. The partial pressure vs time for the Earth's history is presented in Figure 10.15. The Earth's (total) atmospheric pressure is the summation of the partial pressures of all the component gases. Knowing the total atmospheric pressure of a planet, one can then use the following equation to compute the atmospheric temperature (This equation was developed in Chapter 2),

$$T = b^{\alpha}\left[\frac{S(1-A)}{\sigma\left\{4\left(\frac{\frac{\pi}{2}-\psi}{\frac{\pi}{2}}\right)+2\left(\frac{\psi}{\frac{\pi}{2}}\right)\left(\frac{2}{1+\cos\psi}\right)\right\}}\right]^{\frac{1}{4}}\left(\frac{p_t}{p_o}\right)^{\alpha} \quad \text{(Eq. 2.25)}$$

where S is the solar constant at the distance of Earth from the Sun, σ is the Stefan-Boltzmann constant; A is the planet's albedo (for the Earth $A \approx 0.3$); ψ is the precession angle; p_t is the atmospheric pressure; p_o is the atmospheric pressure at sea level, α is the adiabatic exponent + (for the Earth, $\alpha = 0.1905$). and the coefficient b is defined by the following expression:

$$b = \frac{1}{(1-A)^{\frac{1}{4\pi}}} \quad (10.10)$$

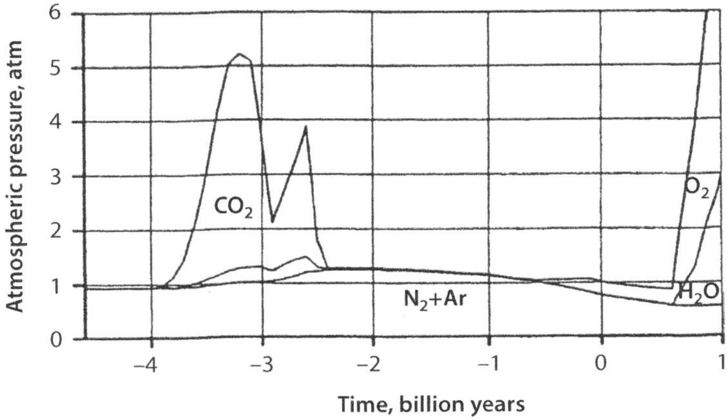

Figure 10.15 Variation of the atmospheric composition and pressure for the Earth over time. (Modified after Sorokhtin, 2005.)

The coefficient for today's atmosphere on the Earth is $ba = 1.093$. Although the value of b does not change if the composition of atmosphere changes the value of ba changes with the value of adiabatic constant a. if the composition of the planet's (Earth's) atmosphere changes.

Eq. 2.25 can be used for comparison of temperature characteristics for the same planet with various gas compositions in its atmosphere. Looking at an extreme case, if one assumes that the existing nitrogen-oxygen atmosphere of the Earth is replaced entirely by a carbon dioxide atmosphere, with the same atmospheric pressure of 1 atm and adiabatic exponent $a = 0.1428$, then the value of $ba = 1.597^{0.1428} = 1.069$ and the near-surface temperature of the Earth would decline to 281.6 K, i.e., the atmosphere would then cool by 6.4 °C. This fact again exposes the statement that *adding carbon dioxide to our atmosphere would warm the atmosphere* is false. Many who support global warming caused by man adding carbon dioxide to the atmosphere do not understand this. Scientific evidence shows that ***as the content of the carbon dioxide in the atmosphere increases, the atmospheric temperature will decrease if the total atmospheric pressure remains the same.***

Another example is the hypothetical situation where the Earth's nitrogen-oxygen atmosphere is totally replaced by a carbon dioxide atmosphere and then constructing the distribution of temperature (at the same atmospheric pressure) for the corresponding elevation. The temperature will be lower than that for an equivalent nitrogen-oxygen atmosphere of the Earth.

Figure 10.16 displays the atmospheric curves showing pressure versus temperature for the Venus troposphere consisting of a carbon dioxide (*Curve 1*) and a hypothetical Venus atmosphere of nitrogen-oxygen (*Curve 2*). In Figure 10.16, *Curve 1*, the graph of temperature distribution for a 100% carbon dioxide troposphere lies below the graph of the nitrogen-oxygen, *Curve 2*, for all pressures (elevations). Correspondingly, the near-surface temperature for the carbon dioxide troposphere occurs 6.4 °C lower than that for the nitrogen--oxygen atmosphere and not considerably higher, as some scientists still, erroneously, continue to believe. These curves demonstrate that increasing the percentage of carbon dioxide in the Earth's atmosphere can only decrease (cool) the temperature of the Earth's troposphere. Insignificant changes in the partial-pressure of carbon dioxide (few hundred ppm) would not have a noteworthy influence on the ability of the troposphere to absorb energy.

By analogy, if one assumes that the existing carbon dioxide atmosphere of Venus was entirely replaced by a nitrogen-oxygen atmosphere, at the same pressure of 90.9 atm, then its surface temperature would increase from 735 to 796 K (462 to 523 °C). Thus, one can once again see, that for

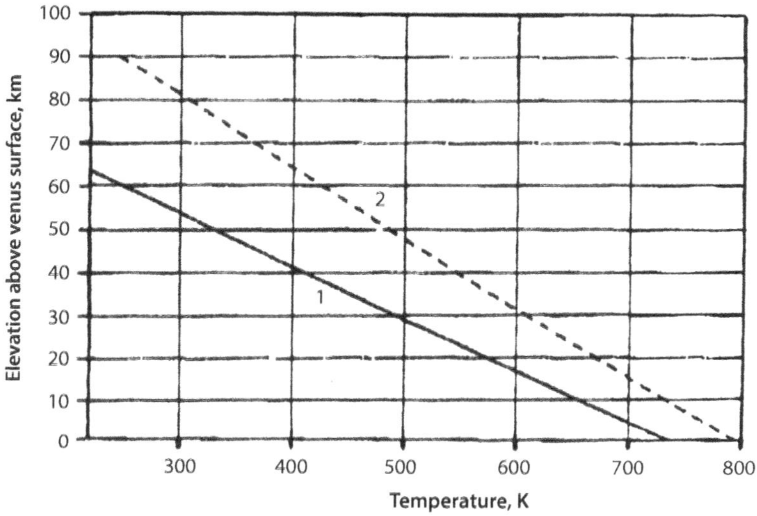

Figure 10.16 Averaged temperature distributions in the troposphere of Venus: *Curve 1* – Today's carbon dioxide troposphere and *Curve 2* – hypothetical model of a nitrogen-oxygen atmosphere at the same conditions. (After Sorokhtin *et al.*, 2007, figure 7.10, p. 274.)

an oxygen-nitrogen atmosphere, increasing the carbon dioxide content, under the same other conditions, would lead to cooling of the entire planet's atmosphere.

Using Eq. 10.9, after differentiation and transition to finite differences (see also Khilyuk and Chilingar 2003), and if the atmospheric pressure, $p_s \approx 1$ atm, one obtains the following equation:

$$\Delta T_s \approx T\alpha(\Delta p_s) \quad \text{(Eq. 10.11)}$$

where ΔT_s is the change in temperature at sea level attributed to the corresponding change in atmospheric pressure $\Delta p_{s,}$ the average Earth's temperature $T = 288$ K and the adiabatic exponent $\alpha = 0.1905$. For instance, under the doubled carbon dioxide concentration in the Earth's atmosphere from 0.046 to 0.092 mass % (as anticipated by the year 2100), the pressure increase, Δp_s, would reach 0.46 mbar. Using Eq. 10.11, $\Delta T_s \approx +0.025$°C. This temperature rise is not associated with the change in the atmosphere's composition, but due to the increase in atmospheric pressure. Thus, the anthropogenic carbon dioxide releases into the atmosphere have no practical influence on the greenhouse effect in the atmosphere.

According to Henry's law, most of the carbon dioxide released into the atmosphere is dissolved within the oceanic water, and upon hydration of the oceanic crust where it can be bound in the carbonates (some CO_2 is taken up by plants). Part of the atmospheric oxygen, together with carbon, is also fixed in carbonates.

Upon hydration of oceanic crust rocks, a portion of carbon dioxide is reduced to methane. Today, due to the formation of carbonates and methane generation, 2.3×10^8 tons/year of carbon dioxide are removed from the ocean and, therefore, removed from the atmosphere. The potential of this CO_2 removal mechanism, however, is much higher. Although the period of time for this geochemical cycle may be over 100 years, the effect is cumulative.

Together with anthropogenic carbon dioxide, some oxygen is also removed from the atmosphere. Based on the carbon dioxide molecular stoichiometry, almost 2.3 g of oxygen is removed from the atmosphere for each gram of carbon. Provided the ocean and vegetation absorbs all excess carbon dioxide after the year 2100, this will likely result in a decline of atmospheric pressure by $\approx$0.34 mbar and, therefore, results in an additional climatic cooling by $-8.2 \times 10^{-3}\,K \approx -0.008°C$. It appears that the life activity of plants could almost completely restore the equilibrium distorted by humans through an accelerated biomass growth.

Historic Effect of Anthropogenic Carbon Dioxide

Today, a major environmental concern among some is the production and burning of hydrocarbons and the subsequent release of the carbon dioxide to the atmosphere. It is thought by these individuals, without scientific evidence or understanding of the tropospheric thermal dynamic heat exchange, that humans are putting carbon (in the form of carbon dioxide or methane) into the atmosphere and that this anthropogenic carbon will be responsible for *global warming*. The prior discussion theoretically proves that changes of carbon dioxide in the Earth's atmosphere by addition of a greater amount of carbon dioxide will cool, not warm, the troposphere. Looking at the Earth's historical temperature charts for confirmation of the theoretical evidence, Figure 10.14 helps to confirm that recent minor changes in tropospheric carbon dioxide concentration has neither cooled nor warmed the atmosphere. The significant rate of increase in carbon dioxide concentration (slope of line) has not affected the temperature cycle.

As stated in Chapter 1, the concept of carbon dioxide warming the atmosphere began with Syante Arrhenius, a Swedish scientist, who was the *first* to claim that fossil fuel combustion may result in enhanced global warming

(Maslin, 2004; in: Enzler, 2015). He proposed that there was a relationship between atmospheric carbon dioxide concentrations in the atmosphere and global temperature. Since then, the concept has been taken as obvious by many without *verification* (Budyko, 1997; Global warming, 1993; Greenhouse effect, 1989).

The arguments for, against and why global warming occurs, have grown within the scientific community since Syante's proposal, possibly due to not fully understanding how heat is transferred within the troposphere. In the upper atmosphere, where there are fewer molecules and a pressure less than 0.1 atm, radiation is the primary source of heat transfer. In the troposphere (the minimum pressure is 0.3 atm), where there are many more molecules per unit volume, the primary mechanism for heat transfer is conduction (see Chapter 2 and 3). The effect of carbon dioxide concentration in the atmosphere is better understood today because of the studies of the historic-atmosphere found in ice cores drilled in the Antarctica and Greenland ice sheets. These studies show a natural cycling of the Earth's temperature, some fluctuating for periods of ≈100,000 years/cycle, as ancient temperatures were directly measured in the ice-sheet-cores, for the last million years. A temperature chart from Antarctica (Figure 10.17) shows cyclic periods of temperature (*Curve a*) and percentage of carbon

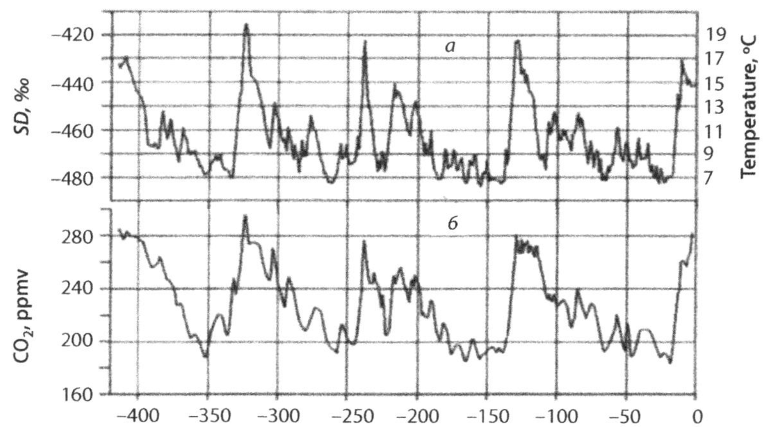

Figure 10.17 *Curve a* – Correlation between the isotopic air temperature and time. *Curve b* – Correlation between the atmospheric carbon dioxide concentration (ppmv) and time over the past 420,000 years at the Antarctic station Vostock, Antarctica. The CO_2 concentration and temperature data were derived from ice cores of a well drilled at that station to a total depth of 3,623 m. The temperature scale is the authors' interpretation of the data. (Data provided by V. M. Kotlyakov.)

dioxide content (*Curve b*) of approximately 100,000 years/cycle for the past 400,000 years. It should be further pointed out that the carbon dioxide curve peaks occur after temperature increases in the tropospheric temperature chart by 400 to 1,000 years, as determined by many historic atmosphere studies in Antarctica and Greenland.

Of importance is the fact that the atmospheric carbon dioxide content (*Curve b*), follows the temperature (*Curve a*). This relationship is common to the many ice core studies done in Greenland and Antarctica, although the period of time that the carbon dioxide curve follows the temperature cure varies between 400 and 1,000 years. The only likely conclusion is that the rise in measured temperature results in an increase of the atmospheric carbon dioxide content (see Chapter 15 for a detailed description).

It has been proposed by several investigators, including Robertson and Chilingar (2017), that this phenomenon of increasing carbon dioxide is the result of rising ocean water temperature (rising temperature reduces the carbon dioxide gas solubility in the ocean) and forcing the dissolved carbon dioxide in the ocean water (reservoir) into the atmosphere (reservoir) (see Robertson and Chilingar, 2017, Chapter 4).

Today, some investigators, ignoring the Earth's historic temperature charts, have erroneously suggested that Earth's global temperature is currently warming. However, Monctkon (2015) in a review of Earth's lower atmosphere temperature data, obtained by satellite, has noted little evidence of either warming or cooling for the past 17 years. Bastasch (2015) in a similar study, after a review of the data and collection methods from America's most advanced climate monitoring system, the National Oceanic and Atmospheric Administration (NOAA), also found that the U.S. has undergone a mild cooling trend over the last decade, despite recent statements by some scientists claiming that *global warming* has been accelerated worldwide. Figure 10.18 was prepared by the U.S. Climate Reference Network, developed by NOAA gathered over the past ten years to provide *high-quality* climate data. NOAA's network consists of 114 stations across the U.S.

A review of temperature deviation data over the past 2,000 years reveals that the Earth's temperatures have been mildly cyclic over this period and that we are currently in a *modern warm period* (Figure 10.19). An examination of this figure indicates a cycling of temperature and an overall gentle cooling period from AD 1000 to 1700 and that the temperatures are: (1) cyclic and (2) today's warm temperatures are cooler than the earlier medieval warm period.

Evans (2010) studied the relationship between global atmospheric carbon dioxide content, human generated carbon dioxide emmissions versus

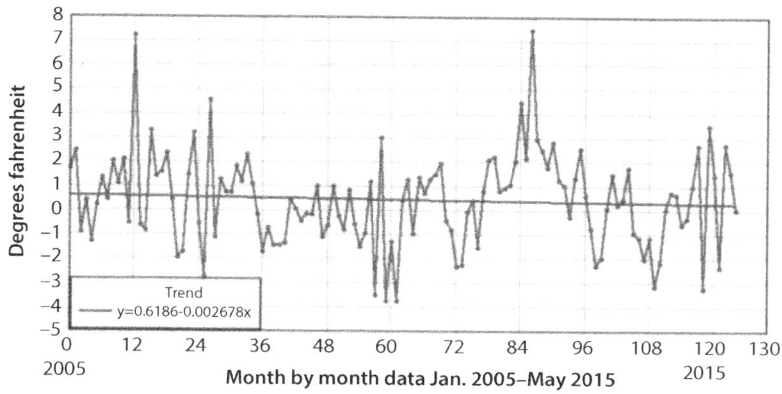

Figure 10.18 Contiguous U. S. average temperature anomaly (°F) from 2005 to 2015. (After Watts, 2015; source NOAA U.S. Climate Reference Network, USCRN.)

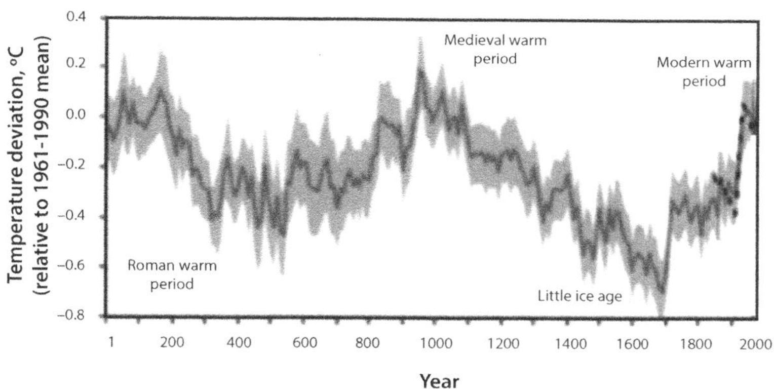

Figure 10.19 A reconstruction of temperature deviation, to the 1961 - 1900 mean temperature for the Earth's Northern Hemisphere during the last two millennia. The fact that the current warm period is not exceptional is clearly shown. (Modified after Geografiska Annaler, 2010.)

global temperature from A.D. 16 to 2010. He noted that the significant increase in human carbon dioxide emissions shown in Figure 10.14, appeared to have had little or no effect on the cyclic global temperatures over this period. In fact, there is no justification for claiming that increased carbon dioxide content due to human activity can increase the global temperature. Examining only the supposed human-generated carbon emissions and then comparing the data to the global temperature charts, there is no significant relationship between the increase of man-made emissions of carbon content, in the atmosphere and the temperature increase (Figure 10.14). This also assumes that the increase in global temperature is due to the increased CO_2 emissions and no other factors, e.g., the Sun/Earth relationship and volcanism. Further supporting the notion that CO_2 emissions are not a factor in raising global temperatures, Evans (2010) noted that this figure shows that today's global warming trend was initiated prior to 1700; however, the additional man-made emissions had negligible effect on temperatures from 1850 to 1940, indicating that there is no possibility that humans initiated global warming. Evans (2010) data also questions if there has been any significant global warming since 1998. Since a quarter of all emissions caused by human activity have occurred in the past 12 years, why have we not seen a greater increase in temperature? Jones (2009) stated that perhaps man-made emissions of CO_2 have merely aggravated the current global warming trend. He also pointed out that 85% of all such emissions have occurred after 1945. The post-World War II industrialization greatly accelerated human emissions as shown in Figure 10.17.

Hiebe (2009) has suggested that we are currently in an ice age climate period. "However, *for the last 10,000 years we have enjoyed a warm but temporary interglacial vacation.*" This temperature record is chronicled in the geological records, e.g., (1) ocean sediments and ice cores from permanent glaciers that have formed in the last 750,000 years and (2) interglacial periods that occur at ≈100,000-year intervals. Hiebe (2009) also suggested that these cycles have been occurring for the past 2 to 4 MY although the overall Earth's temperature, as investigated by Robertson and Chilingar (2017) and by several other investigators utilizing many independent methods, has been continuously cooling for the past 65 MY (see Figures 1.3, 1.4, and 1.5).

The above discussion leads to only one conclusion: the theoretical effect of increasing the content of carbon dioxide emissions doesn't lead to warming of the troposphere and this theory is strongly confirmed by historical evidence.

11

Ozone in the Earth's Atmosphere

Properties of Ozone

Ozone (O_3) is a pale bluish colored gas with a pungent odor, composed of three atoms of oxygen. Normal oxygen (O_2) is a gas that humans breathe, which has two oxygen atoms and is colorless, odorless and represents about 21% of the Earth's atmosphere.

The ozone molecule is very rare in the Earth's atmosphere, averaging about three molecules of ozone for every 10 million air molecules. About 90% of the ozone is found in the stratosphere where it is concentrated in a layer, about 16 to 40 kilometers (6 to 31 miles) above the Earth's surface (Figure 11.1). Ozone concentrations in this layer are typically only 1 to 10 parts of ozone per 1 million parts of atmosphere (air), compared with about 210,000 parts of oxygen per 1 million parts of air. The other 10% of the ozone is found in the troposphere with concentrations ranging from ≈0.02 to 0.3 parts per million. Welch (2017) noted that the ozone layer extends over the entire globe with some variation in altitude and thickness, being thinner about the south pole.

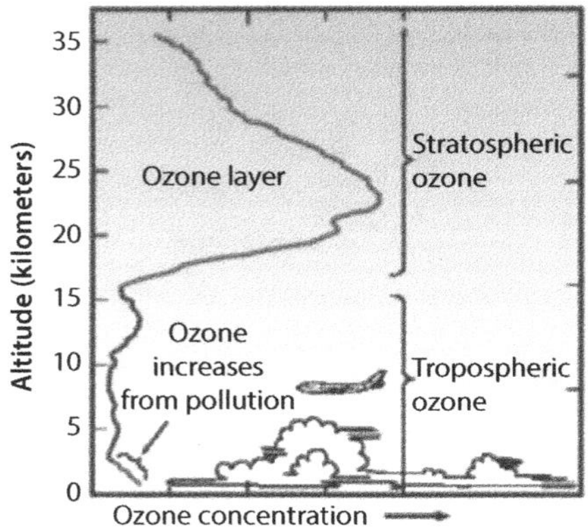

Figure 11.1 Ozone concentration in the Earth's troposphere and stratosphere. (After Ozone, 2014.)

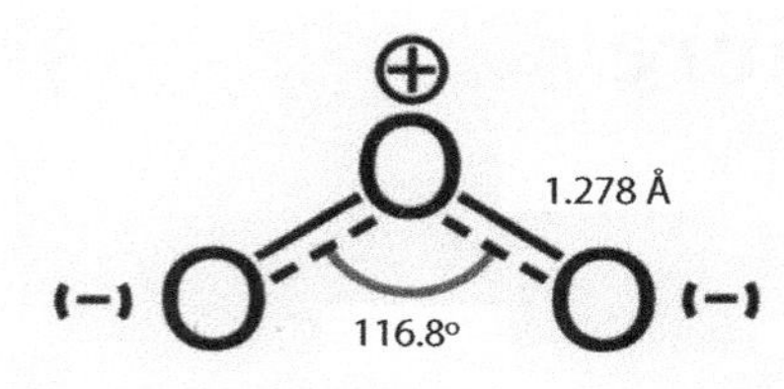

Figure 11.2 Diagram of ozone molecule. (After Boundless, 2017.)

Ozone is a bent molecule (Figure 11.2), with C_{2v} symmetry, similar to the water molecule. According to Wikipedia (2017): (1) the O–O distances are 127.2 pm (1.272 Å); (2) the O–O–O angle is 116.78°; (3) the central atom is sp^2 hybridized with one lone pair; (4) ozone is a polar molecule with a dipole moment of 0.53 D; and (5) ozone is unstable and very reactive.

Ozone is slightly soluble in water, and much more soluble in inert non-polar solvents such as carbon tetrachloride (CCl_4) or fluorocarbons, where it forms a blue solution.

Ozone Layer as the "Earth's Shield"

According to Khromov and Petrosyan (2001), ozone absorbs the solar radiation with the wave lengths ranging from 0.15 to 0.29 *m*m of the solar spectrum, providing a shield that protects life on the Earth's surface. Whereas both oxygen and ozone together absorb 95 to 99.9% of the Sun's ultraviolet radiation, ozone effectively absorbs most of the ultraviolet light known

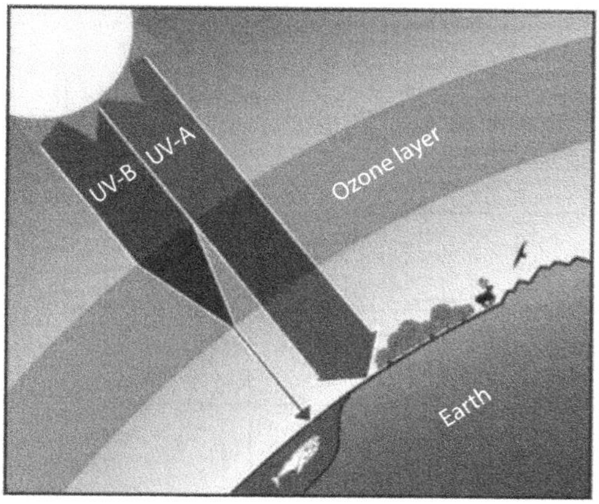

Figure 11.3 UV protection by the ozone layer. UV represents the Sun's ultraviolet (UV) radiation which is made up of three types of rays -- ultraviolet A (UVA), ultraviolet B (UVB), and ultraviolet C (UVC). UVC is the most dangerous type of ultraviolet light but cannot penetrate Earth's protective ozone layer. (After Welch, 2017.)

as UV-A and UV-B, which can cause biological damage (Figure 11.3). Ultraviolet UV-B is absorbed by the ozone layer and little of it reaches the Earth's surface. The term *shield* as a description of the layer of ozone in the stratosphere is a bit misleading, because the molecules do not form an impermeable sphere around the earth. But they do block out, or absorb, about 50% of the incoming Sun's rays (Welch, 2017).

Ozone is a very reactive gas and considered unstable. It continuously breaks apart into its oxygen atoms and then reforms as an ozone molecule. As an ozone molecule, it doesn't last very long. Although the *shield* (layer of ozone molecules) changes constantly, the atmospheric chemical processes maintain a dynamic equilibrium that keep the overall amount of ozone in the atmosphere constant (Welch, 2017).

Ozone is constantly being formed by the reaction to the Sun's ultraviolet radiation on oxygen molecules in the Earth's atmosphere. The UV light rays split the gas molecules (O_2) apart by breaking the bonds between the atoms. The reactive free oxygen atom then collides with another oxygen gas molecule to form an ozone molecule. Because ozone is unstable, ultraviolet light supplies the energy that quickly breaks it up, and the process begins again (Welch, 2017).

Atmospheric Gases Ability to Absorb Energy

It is often assumed that the energy absorption (UV portion of the solar radiation) occurs in the stratosphere due to the ozone (a triatomic oxygen)

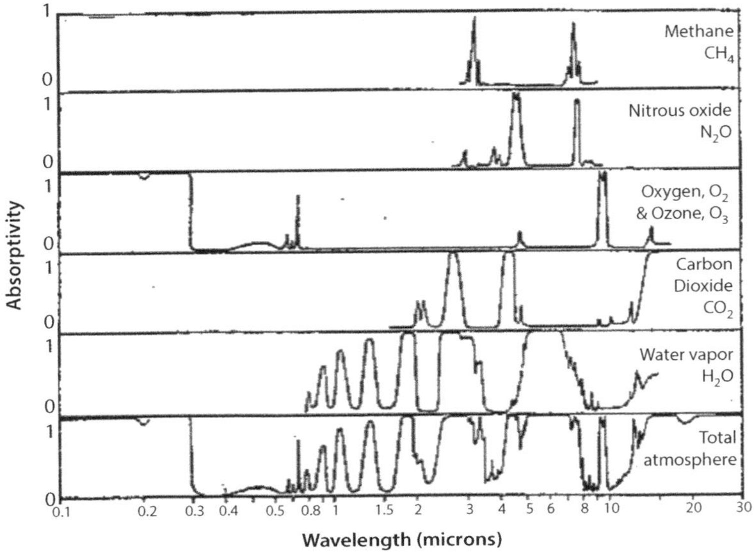

Figure 11.4 Wave length radiation absorptivity for various gases of the atmosphere. An absorptivity of zero means no absorption, while a value of one means complete absorption. The dominant gas absorbers for infrared radiation are water vapor and carbon dioxide, oxygen and ozone. (After J. N. Howard, 1959: Proc. I.R.E. 47, 1459; and R. M. Goody and G.D. Robinson, 1951: Quart. H. Roy. Meteorol. Soc. 77, (153); in: http://www.climatesight.org/2011/03/30/ozone-depleytion-and-climate-change/.)

forming reaction and that the stratospheric ozone is the primary medium that absorbs the solar UV radiation. Khromov and Petrosyants (2001) and Real Science (2014) note that ozone absorbs the solar radiation with the wavelengths of 0.15- to 0.29-mkm, which is, in the UV portion of the solar spectrum. However, there are several other atmospheric gases that also absorb radiation. Figure 11.4 lists the ability of several atmospheric gases that absorb energy as a function of the wave length radiation.

The absorption of solar UV rays within the Earth's atmospheric layers is mostly associated with the photochemical dissociation of the oxygen molecules accompanied by absorption of the energy of the hard-solar radiation. The absorption of the Sun's UV radiation occurs within the stratosphere and mesosphere due to oxygen molecule dissociation into the atoms:

$$O_2 + hv \rightarrow 2O \qquad \text{(Eq. 11.1)}$$

where hv is the energy of the Sun's UV radiation, h (6.626×10^{-27} erg · s) is the Planck's constant, v is the frequency of the absorbed UV electromagnetic fluctuations. For Eq. 11.1, $hv = 5.12$ ev; v^{3} 1.24×10^{15} Hz at the wave

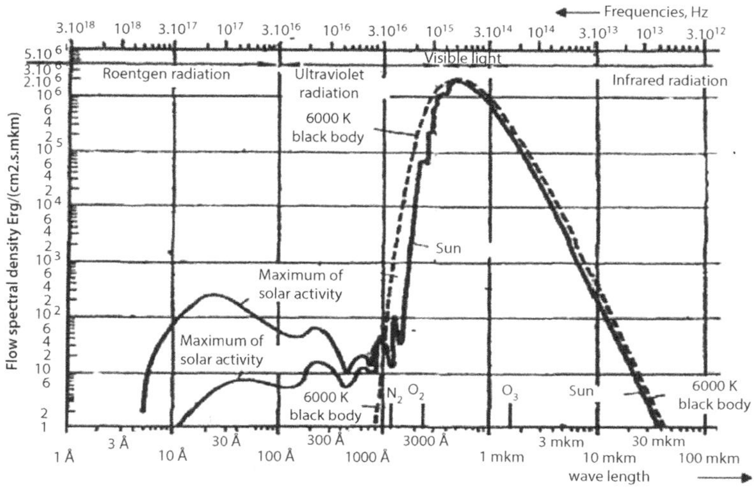

Figure 11.5 Current solar radiation spectrum (solid line) compared to the spectrum of absolute black body at 6000 K (dashed line). Symbols N2, O2, and O3 indicate the spectrum lines (Fraunhofer lines) of nitrogen, oxygen, and ozone molecules corresponding to the dissociation energy of these molecules. (After Sorokhtin *et al.*, 2011, figure 12.46, p. 463.)

length $\lambda \leq 2420$ Å. The reverse reaction of oxygen atoms joining into a molecule occurs with the energy (heat) released

$$O + O \rightarrow O_2 + 118 \text{ kcal/mol}. \qquad \text{(Eq. 11.2)}$$

Similar reactions occur with nitrogen gas in the atmosphere:

$$N_2 + hv \rightarrow 2N \text{ and } N + N \rightarrow N_2 + Q. \qquad \text{(Eq. 11.3)}$$

The nitrogen effect may be disregarded, as the solar hard radiation intensity at the frequency of the nitrogen molecules dissociation, is about four orders of magnitude lower than at the oxygen dissociation frequency (see Figure 11.5).

The water vapor and carbon dioxide percent concentrations in today's stratosphere are too low to be considered. A different situation occurred in the Archaean time with an atmosphere of a denser carbon dioxide content and pressure that reached 5 atm. Because of this, in the Archaean time the absorption of the solar UV radiation was at least as intense under the reaction for CO_2 dissociation:

$$CO_2 + hv \rightarrow CO + O. \qquad \text{(Eq. 11.4)}$$

During the Archaean time, like in the Proterozoic, warm climate with its nitrogen oxygen-devoid atmosphere, the UV radiation was apparently also absorbed in the troposphere by water vapor (humidity) dissociation within the cloud cover according to the following equation,

$$OH + O \rightarrow H_2O + Q \qquad \text{(Eq. 11.5)}$$

Because of the atmospheric radiation-convection balance, the average temperature at the surface of the Earth is positive +15 °C, although its fluctuations in different climatic zones may reach 150 °C.

$$O_2 + hv \rightarrow 2O \qquad \text{(Eq. 11.6)}$$

$$H_2O + hv \rightarrow OH + O \qquad \text{(Eq. 11.7)}$$

Today, absorption of the Sun's UV radiation in the tenuous air of the troposphere, stratosphere and mesosphere is caused mostly by the photochemical dissociation of oxygen and water molecules, which is accompanied by the absorption of the Sun's hard radiation.

Solar energy absorption occurs with ozone dissociation. Under the conservation of energy law, the amount of the absorbed energy is the same as in its formation, that is, $hv = 24$ kcal/mol:

$$O_2 + O \rightarrow O_3 + hv, \qquad \text{(Eq. 11.8)}$$

$$O_3 + hv \rightarrow O_2 + O \qquad \text{(Eq. 11.9)}$$

The heat release in the ozone molecule formation results in the heating of the tenuous air mass in the stratosphere and mesosphere which is clearly seen on a temperature profile of the atmospheric layers (see Figure 11.1).

The formation of ozone can only occur with the heat release under an exothermal reaction and not with the absorption of the solar radiation energy as is still sometimes believed:

$$O_2 + O \rightarrow O_3 + 24 \text{ kcal/mol} \qquad \text{(Eq. 11.10)}$$

The absorption of the IR solar radiation will only occur with ozone dissociation. Under the energy conservation law, at ozone decomposition into the source components, the same energy amount is absorbed as in its formation, i.e., $hv = 24$ kcal/mol:

$$O_3 + 24 \text{ kcal/mol} \rightarrow O_2 + O \qquad \text{(Eq. 11.11)}$$

The molecular dissociation energy is usually measured in electron-volts, 1 ev = 1.602×10^{-12} erg or 23.045 kcal/mol. The dissociation of one oxygen molecule is accompanied by the energy absorption, E, about 5.12 ev or 8.203×10^{-12} erg, ≈ 118 kcal/ mol, and that of carbon dioxide, by the energy absorption 5.45 ev or 8.731×10^{-12} erg, ≈ 125.6 kcal/ mol, at water dissociation, 5.11 ev or 8.186×10^{-12} erg, ≈ 117.8 kcal/mol of energy is absorbed, and at ozone dissociation, about 1.04 ev or 1.67×10^{-12} erg, ≈ 24 kcal/mol more. The critical frequencies at absorption of the electromagnetic waves of these reactions are apparently, $v = E/h$, where, E is given in ergs, and the critical wave lengths $l = c/v$ cm, where $c = 2.998 \times 10^{10}$ cm/s is the light velocity At the oxygen dissociation, the absorbed frequencies are higher than 1.238×10^{15} Hz, and the wave lengths are shorter than 2.422×10^{-5} cm; at carbon dioxide dissociation, the frequencies higher than 1.318×10^{15} Hz are absorbed, and the wave lengths shorter than 2.275×10^{-5} cm; at the water molecule dissociation, the frequencies higher than 1.235×1015 Hz are absorbed, and the wave lengths shorter than 2.427×10^{-5} cm. Therefore, the dissociation of oxygen, carbon dioxide, and water molecules must result in the absorption of the solar UV radiation, as all dissociation spectra of these molecules are within the UV portion of the spectrum. The boundary of the visible and UV radiation in the solar spectrum is at the frequency $v \approx 1 \times 10^{15}$ Hz ($l = 3 \times 10^{-5}$ cm). The restoration of oxygen, carbon dioxide, and water molecules apparently results in the heat release Q and is exactly equal to the dissociation energy of these molecules. The described reactions convert the energy of the solar UV radiation into heat.

The frequency of the solar radiation absorption by ozone is $v \approx 0.252 \times 10^{15}$ Hz or $\lambda \leq 1.19 \times 10^{-4}$ cm. This is within the IR portion of the spectrum; therefore, ozone can only absorb the short-wave portion of the heat waves and the adjacent low-frequency area of the visible solar radiation but not the UV (see Figure 11.5).

The above estimates and reasoning lead to a conclusion that was important for life on the Earth, beginning in Archaean time, at the very moment of the appearance of life on the Earth, the Earth's surface was never subjected to intense solar UV radiation which allowed life to evolve directly on the Earth's surface. This is supported by the emergence in Early Proterozoic time of red-colored weathered crusts formed because of iron-oxidizing bacteria metabolism (Anatolyeva, 1978). The similar Archaean time crusts apparently were not preserved due to an active erosion of Archaean time continents.

Only the young Earth in Katarchaean (4.6 to 4 BY ago) was totally devoid of a hydrosphere and chemically active atmosphere, so it was actively irradiated by the intense UV radiation, which completely prohibited the emergence of life on the Earth prior to the Archaean time.

Looking at today's conditions and to the dissociation of only atmospheric oxygen, if after the dissociation of oxygen molecules, all released oxygen atoms recombine, the released heat may be determined as follows:

$$Q = \frac{h\nu N_A}{K_{td}} = \nu\left[\frac{hN_A}{K_{td}}\right] = 9.531\times10^{-11}\nu \qquad \text{(Eq. 11.12)}$$

Here, the Avogadro's number, $N_A = 6.023 \times 10^{23}$ and $K_{td} = 4.187 \times 10^7$ erg/cal is the thermodynamic conversion factor from ergs to calories. Then, for oxygen,

$$(1.24 \times 10^{15})(9.531 \times 10^{-11}) = 118{,}184 \text{ cal/mol} \approx 118.2 \text{ kcal/rnol}$$

and for ozone:

$$(0.252 \times 10^{15})(9.531 \times 10^{-11}) = 24{,}018 \text{ cal/mol} \approx 24 \text{ kcal/mole.}$$

Therefore, the inverse recombination reactions of the earlier dissociated oxygen molecules into air molecules, transform the energy of the solar UV radiation into the heat energy, and is spent for the heating of a thin layer of stratospheric or mesospheric air at a height of 15 to 80 km. This can be clearly seen on the temperature profile of these atmospheric layers (see Figure 11.6).

The confusion with the UV radiation absorption by ozone is likely associated with the assumptions in the calculations of the ozone molecules' complete dissociation into three oxygen atoms at once, with the absorption of energy of 24.02 + 118.18 = 142.2 kcal/mol. This process is a two-step one: (1) Initially, ozone releases only one atom (Nekrasov, 1973), with the absorption of 24.02 kcal/mol, and (2) then the remaining molecules mix with the atmospheric oxygen and dissociates with the absorption of 118.18 kcal/mol, but only if affected by the other photons as is shown under reaction in Eq. 11.1.

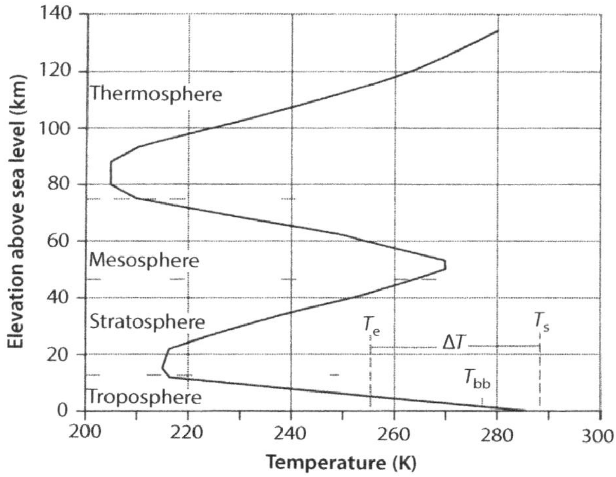

Figure 11.6 Temperature distribution in the Earth's atmosphere (Atmosphere of Earth, 1988). T_e is the Earth's effective temperature: T_s is the Earth's average temperature normalized for sea level. DT is the greenhouse effect value. T_{bb} is the absolute black-body temperature at the distance of the Earth from the Sun. The tropopause temperature ranges from 190 to 220 K. (After Sorokhtin *et al.*, 2011, figure 13.2, p. 473.)

The atmospheric pressure at the elevation h is approximately

$$p_h = p_o e^{\left(h\frac{g\mu}{RT}\right)} \qquad \text{(Eq. 11.13)}$$

where the pressure at the Earth surface is $p_0 = 1.013$ mbar; the gravity acceleration is $g = 981$ cm/s^2; the air mixture average molar weight is $m = 28.89$; the gas constant is R = 8.314×10^7 erg/mol · deg; the approximate temperature at the stratosphere/mesosphere boundary is $T \approx 270$ K, that is, at the elevation of the maximum stratosphere heating of about 50 krn (Figure 10.6). Then pressure at that elevation is:

$$p_{50} \approx 1.013e^{\left[5\times10^6\left(\frac{981\times28.89}{8.314\times10^7\times270}\right)\right]} = 1.837 \text{ mbar}$$

At the adiabatic air temperature distribution, within the atmosphere, its undisturbed value at the elevation, h, (Sorokhtin, 2011) is:

$$T_h = 288\left(\frac{p_h}{p_o}\right)^{\alpha} \qquad \text{(Eq. 11.14)}$$

where $a = 0.1905$ is the atmospheric adiabatic exponent and 288 K is the average Earth temperature at sea level. Then the adiabatic temperature at the elevation of about 50 km will be:

$$T_{50} = 288\left(\frac{1.837}{1013}\right)^{0.1905} \approx 87\text{ K}$$

At recombination of the oxygen atoms into the molecules under reaction (Eq. 11.2), additional stratosphere heating occurs. If the fraction of the dissociated oxygen molecules was only 0.3% ($S = 0.003$), then the heat emitted in today's atmosphere is:

$$\Delta T = S\frac{Q}{c_p} = 0.003\left(\frac{118{,}000}{1.962}\right)180\text{ K}$$

where the air heat capacity, $c_p \approx 1.962$ cal/mol · deg, at 87 K the resulting temperature at the 50 km elevation will be $(87 + 180) = 267 \approx 270$ K as shown in Figure 11.6.

At elevations below 50 km, the UV radiation intensity declines. Thus, the relative number of the molecule dissociation events significantly decrease, and so does the stratosphere heating. At the same time, at greater elevations the decrease in the number of the molecule dissociation events is in proportion with the exponential air pressure decline in the same direction.

The ozone's mass content in the atmosphere is about 3.6×10^{-5} %. Thus, if every single ozone atom takes part in the absorption of the solar IR radiation, a similar calculation of the stratosphere and mesosphere heating will be only 6.7 K. Therefore, the ozone formation may result in a significantly smaller, by the factor of 30, atmosphere heating than the observed. This confirms the hypothesis according to which the responsibility for the absorption of the solar UV radiation lies not with ozone but with the dissociation process of oxygen molecules into atoms.

Figure 11.7 describes the variation of the ozone concentration for the Earth's atmosphere versus time. The percentage ozone appears to be related to sunspot activity. In times of greater sunspot activity, there is a lower ozone content in the atmosphere. In general, this figure shows that the level of ozone has decreased from prior years.

The emergence of the oxygen atoms in the transition layer between the stratosphere and mesosphere results in the occurrence in that layer of a noticeable, oxidizing environment, which leads for instance, to the oxidizing of any methane which happens to get there:

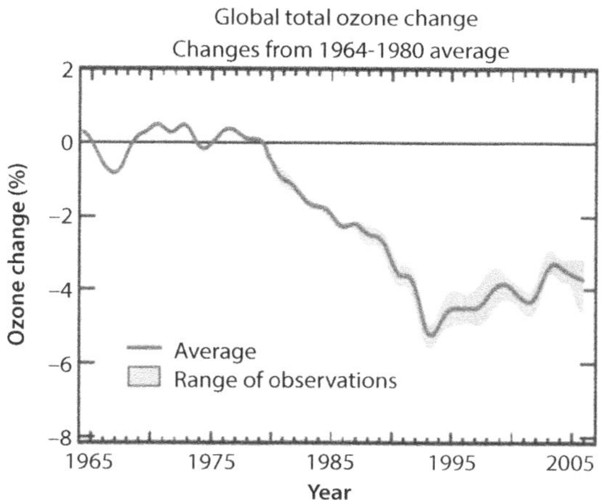

Figure 11.7 Variation in the global ozone concentration with time. It has leveled off after years of decline and has been rising since 1993. (After *http://w.w.w.climatesight.org/2011/03/30/ozone-depletion-and-climate-change/*.)

$$CH_4(\text{methane}) + O \rightarrow HCOH(\text{formalehyde}) + H_2 + 9.8\ \text{kcal/mol} \quad \text{(Eq. 11.15)}$$

$$H_2 + O \rightarrow H_2O + 57.8\ \text{kcal/mol} \quad \text{(Eq. 11.16)}$$

$$HCOH + O \rightarrow CO_2 + H_2 + 55.4\ \text{kcal/mol} \quad \text{(Eq. 11.17)}$$

$$H_2 + O \rightarrow H_2O + 57.8\ \text{kcal/mol} \quad \text{(Eq. 11.18)}$$

These reactions decompose the methane, thus stabilizing its atmospheric concentration. They also contribute to the stratosphere and mesosphere heating, which, however, is insignificant due to the low-atmospheric concentration of methane.

Therefore, ozone dissociation and even more so its formation cannot absorb the solar UV radiation as these reactions occur in the IR radiation domain. The solar UV radiation, with a wavelength shorter than 0.24 mcm ($2400\ \mathring{A}$) is absorbed only by the oxygen dissociation process. Repeated recombination of the oxygen atoms into molecules results in heating the stratosphere and mesosphere. Thus, the ozone layer in the stratosphere is not a protective shield against the UV radiation which is deadly for all living things. The real *shield* is the oxygen. Still, ozone is generated due to merging of the oxygen atoms released upon the absorption of UV radiation with the oxygen molecules in the air. ***The ozone concentration in the stratosphere may be an indication, but not the cause of the absorption of the solar UV radiation.*** This justifies a closer examination of the ozone layer, although the ozone formation appears to be just a consequence and not the cause of the absorption of the solar UV radiation.

The Ozone Hole

In 1984, British Antarctic Survey scientists Farman, Gardiner, and Shanklin recognized the phenomena of the springtime ozone thinning in the polar area, occurring in the lower layer of the stratosphere over Antarctica. Since then, it has also been identified as taking place in other areas of the ozone layer. Their paper was published in *Nature*, May 1985. Their immediate question was what did this hole mean and how did it occur?

It was first suggested that the *ozone hole* in the Antarctic was a product of pollution by carbon dioxide and related to the mechanism of *global warming*, pertaining to the pollution of carbon dioxide and thinning of the ozone layer. Environmentalists began making numerous apocalyptic predictions pertaining to these holes, assuming incorrectly, that these holes are somehow related to atmospheric carbon dioxide content. Research on this topic, since 1986, has shown that the *ozone hole* is not related to global warming or the carbon dioxide pollution, but rather sunspot activity and is not a mechanism of global warming (Maugh II, 1986). The thinning of ozone layer in the polar area over Antarctica is most likely caused by increased sunspot activity, solar, rather than by man-made carbon dioxide as shown in Figure 10.7 (Tallbloke, 2015). While solar sunspot number closely follows Earth's temperature anomaly, the carbon dioxide content of the atmosphere does not. This increased solar activity produces its most dramatic effect over Antarctica because of its long, lightless and cold winters and explains the thinning that has been observed for other sites of the ozone layer around the world.

The ozone holes are in the region of the stratosphere, located primarily in the polar and temperate latitudes, with a reduced (by approximately 20–30%) concentration of ozone. As a rule, these holes tend to emerge during the winter and spring time above areas of stable anticyclones. e.g., Antarctica and Yakutia. In winter for the Antarctic polar region, this thinning is partially due to the uplifting air masses which overflow into the stratosphere (Zueva *et al.*, 2015). As a result, the ozone layer above these areas is rarefied. In summer time, the ozone holes shrink or totally disappear.

This phenomenon was brought to the attention of scientists at the end of the 1950s when the researchers began measuring the ozone concentration in the stratosphere. The first ozone hole was discovered in Antarctica and, shortly after, speculation on the possibility of their anthropogenic origin was prevalent. Freon gas was at that time identified as the most probable culprit for the hole. However, several crucial factors remained unexplained: (1) why the deepest and widest holes in the ozone layer were observed in the Antarctica (the Southern Hemisphere), whereas freon gas was released mostly in the Northern Hemisphere; and (2) during volcanic eruptions, natural freons were released into the atmosphere in significantly greater quantities than the quantities of industrial freons. Energetically and quantitatively, the quantity of industrial freons are insignificant compared to the role played by natural gases (methane, hydrogen, and natural freon) of volcanic origin.

Initially, it was publicized that the shrinking of the ozone layer spelt doom for humanity. There were conflicting reports from several observers

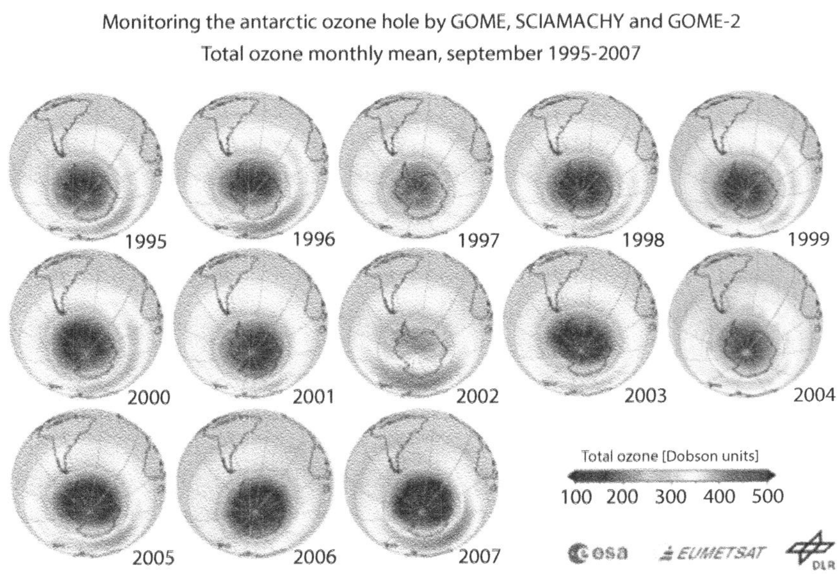

Figure 11.8 The images of the ozone hole display the variability of the size of the hole, each September over many years. The size of the ozone hole relates to the number of sunspots or period of the sunspot cycle and not to the carbon dioxide content of Earth's atmosphere. (After D'Aleo, 2011.) Figure 11-9. Actual Photos of the Antarctic ozone hole (black center) showing two extremes of the hole width. This hole was not visible in 1970, but it was very visible (see dark black area) in 2008. (NASA images courtesy Goddard Space Flight Center, Ozone Processing Team.)

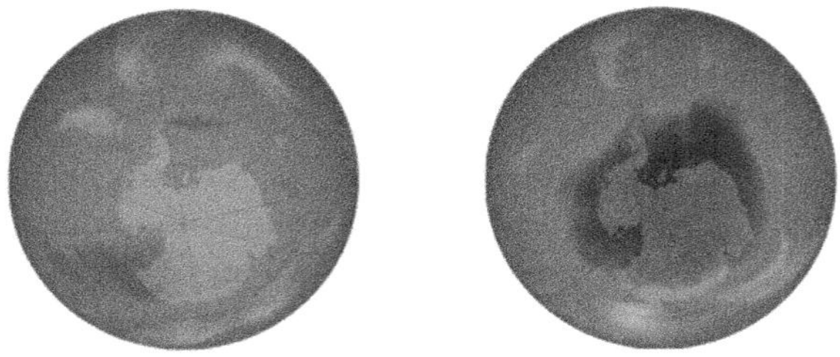

Figure 11.9 Actual Photos of the Antarctic ozone hole (black center) showing two extremes of the hole width. This hole was not visible in 1970, but it was very visible (see dark black area) in 2008. (NASA images courtesy Goddard Space Flight Center, Ozone Processing Team.)

reporting on the shrinking and/or increasing of the total global ozone levels. Satellite pictures of the growth and shrinkage of the hole are available on the internet (Figure 11.8). Ajavon *et al.* (2007) has reported that the global ozone atmospheric concentrations have decreased since 1975 (see Figure 11.7). An examination of this curve indicates a cyclic nature, the thickness of this layer with a minimum value occurring around 1992. Typical satellite photographs of the ozone hole over Antarctica are shown in Figures 11.9, displaying a minimum and maximum width of the ozone hole. A yearly collection of ozone hole photos shows that this hole opens and closes, possibly cyclic with respect to the quantity of energy received from the Sun. In general, the ozone layer thickens when the hole closes. Figure 11.10 shows the hole located over Antarctica in 2011.

The Antarctic ozone hole, was studied by Zueva *et al.* (2015) who recorded the following: the heterogeneous chemical reactions releasing the photochemically active molecular chlorine that plays a key role in the Antarctic stratospheric ozone destruction. One of the principal components in these reactions is the hydrogen chloride (HCl) that forms on the surface of the polar-stratospheric-clouds (PSCs) during the polar nights at temperatures lower than −78 °C. PSCs form mainly on sulfuric acid (H_2SO_4) aerosols and act as condensation nuclei, which are formed from sulfur dioxide (SO_2). However, the cause of HCl and H_2SO_4 high concentrations in the Antarctic stratosphere, leading to considerable springtime ozone depletion, is still not clear. Based on the NCEP/NCAR reanalysis data over the last 35 years and by using the NOAA HYSPLIT trajectory model, shows that the Erebus volcano gas emissions include HCl and SO_2, reaching the Antarctic stratosphere via high-latitude cyclones with the annual average probability, $\overline{P}_{ann}$ of at least ~0.235 (23.5%). Depending

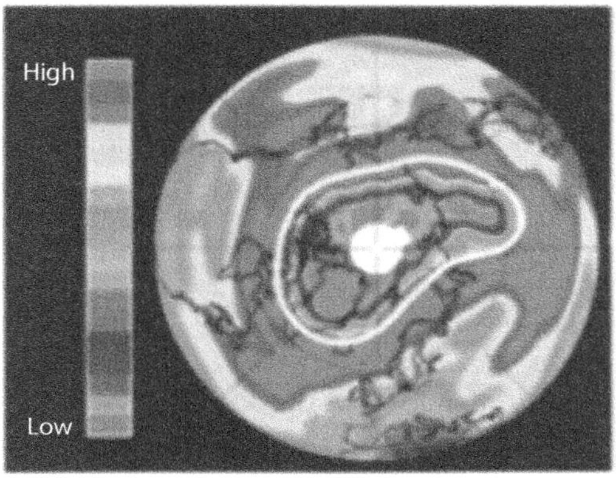

Figure 11.10 Photo of Antarctic area in 2011 displaying continents and low-pressure areas. (After NASA, 2011.)

on Erebus activity, this corresponds to additional annual stratospheric HCl mass of 1.0 to14.3 kilotons (kt) and SO_2 mass of 1.4 to19.7 kt. Thus, Erebus volcano is a natural source of additional stratospheric HCl and SO_2, and hence, the cause of the Antarctic ozone depletion (Zueva *et al.*, 2015).

Ozone holes in the Earth's atmosphere are not limited to just over the Antarctic. On March 15, 2011, a record ozone layer loss was observed, with about half of the ozone present over the Arctic having been destroyed. This change was attributed to the increasingly cold winters in the Arctic stratosphere at an altitude of approximately 20 km (12 mi). By March 25, the area had experienced an 80% loss of ozone. This was the largest loss compared to those observed in previous winters with the possibility that it could become an ozone hole in the stratosphere. Its occurrence would require that the quantities of ozone fall below 200 Dobson units, from the 250 that were recorded over central Siberia (China, 2006).

In 2003, a 2.5 million km^2 with a minimum value of only 190 DU, ozone hole was detected over Tibet (China, 2006). Again, in 2011, an ozone hole appeared over the mountainous regions of Tibet, Xinjiang, Qinghai and the Hindu Kush, along with an unprecedented hole over the Arctic (Figure 11.10). The Tibet hole was far less intense, however, than those over the Arctic or Antarctic. It was the first time that an ozone mini-hole or an extremely low ozone event had been found to occur over the Tibetan Plateau.

Ozone - Methane Reaction

The chemical reaction of methane with ozone occurs as a multi-step sequence:

$$CH_4 + 4O_3 \rightarrow CO_2 + 2H_2O + 4O_2 + 64.96 \text{ kcal/mol} \qquad \text{(Eq.11.19)}$$

and

$$H_2 + O_3 \rightarrow H_2O + O_2 + 25.1 \text{ kcal/mol} \qquad \text{(Eq. 11.20)}$$

where quantity of heat formation of the above constituents is: $CH_4 = 17.89$ kcal/mol; $O_3 = 34.50$ kcal/mol; $O_2 = 2.80$ kcal/mol; $H_2 = 1.00$ kcal/mol; $H_2O = 57.80$ kcal/mol; and $CO_2 = 94.05$ kcal/mol.

The amount of heat generated by the ozone–freon reaction is significantly lower and less than several kcal/mol. Thus, the reaction of ozone destruction by methane releases a far greater amount of energy. According to Eq. 11.19, 4.062 kcal/g of energy is released as a result of the reaction of 1 g of methane with ozone in the stratosphere. As a comparison, burning of 1 g of methane in the air produces 11.98 kcal/g:

$$CH_4 + 2O_2 \rightarrow CO_2 + H_2O + 11.98 \text{ kcal/g} \qquad \text{(Eq. 11.21)}$$

The formation of serpentines in the process of oceanic crust hydration releases about 6 million tons/year of CH_4 and 10 million tons/year of H_2, whereas the technogenic release of freon does not exceed 100,000 tons/year. To this amount one should add many tens of millions of tons of methane and hydrogen entering the atmosphere from the tectonically active regions and the tropical forests. In addition, methane is emitted from the marshes of the northern regions of Canada and Eurasia. The total mass of natural gases entering the Earth's atmosphere reaches hundreds of millions of tons annually. The amount of natural methane, hydrogen, and volcanic Freon entering the atmosphere is almost four orders of magnitude higher than the amount of technogenic freon. The heating effect from the reaction of methane with ozone within the atmosphere is considerably higher than the chemical reaction of freon with ozone.

Kapitza and Gavrilov (1996) showed that variations in ozone concentration in the stratosphere are solely natural and occur with seasonal periodicity. Moreover, they discovered that the ozone concentration in the layers above the equatorial and tropical areas was lower than that within the deepest ozone holes at the near-polar areas. The fluctuations in concentration in the ozone layer are natural and cannot be attributed to human activities.

This leads to the following conclusion: "***the anthropogenic effect on the ozone layer in the stratosphere is negligible in comparison with the effect of natural factors.***" They are approximately three to four orders of magnitude lower.

Concluding Remarks

In summary, the layer of stratospheric ozone is primarily created by ultraviolet (UV) radiation coming from the Sun. The Sun's energy varies, especially over the 11-year sunspot cycle. The length of this cycle is now under study; some researchers are suggesting a longer, 15-year cycle (Tallbloke, 2015). During the high-energy emission phase of this sunspot cycle (see Figure 10.7), more stratospheric ozone is produced, due to the increased UV coming to the Earth. This process increases the average ozone concentration in the stratosphere over the Earth's poles by ≈4%, but when averaged out over the whole Earth, however, the world average ozone increase is only ≈2% (Ozone, 2009). Observations of ozone levels since the 1960s (Figure 11.7) have shown that total atmospheric levels have decreased by 1–2% from 1965 to 2005.

Unusual solar activity can cause the ozone levels in the upper stratosphere to be substantially depleted, but since most of the ozone is in the middle stratosphere, the effect on the total ozone layer is negligible (Ozone, 2009).

As discussed in the prior section, volcanic eruptions can have a diminishing effect on the ozone layer. Zueva *et al.* (2015) has noted that in Antarctica, large volcanic eruptions can inject significant quantities of chlorine, e.g., hydrochloric acid, into the stratosphere where the highest concentrations of ozone are found. However, most of the volcanic eruptions are too weak to reach the stratosphere, around 10 km above the surface (Ozone, 2009). In 1991, Mt. Pinatubo in the Philippines erupted, sending tones of dust and gas high into the atmosphere, which caused global reductions in the ozone layer for two to three years. Thus, large volcanic eruptions may also increase the rate of stratospheric ozone depletion.

12

Evolution of Atmospheric Composition and Pressure

The thickness of the Earth's atmosphere is about 300 miles (480 kilometers). Due to gravity, most of the mass of the total atmosphere lies within a layer ≈8 miles (16 km) thick over the Earth's surface. The lowest layer of the Earth's atmosphere is referred to as the troposphere. As described in the earlier chapters, the Earth's atmosphere composition and pressure have changed over time. The components of today's atmosphere are listed in Table 12.1. Air pressure decreases with altitude because of gravity as shown in Table 12.2. At sea level, the air pressure is ≈14.7 pounds per square inch (1 kilogram per square centimeter). At 10,000 feet (≈3 km), the air pressure is much lower, 10 lb/in^2 or psi (0.7 kg per square cm).

Partial Pressure of Atmospheric Gases

The total pressure of the Earth's atmosphere is a summation of the partial pressures of all the gases that make up the atmosphere. These primarily include nitrogen, oxygen, argon, carbon dioxide, methane and water vapor (see Table 12.1). A plot of the historic evolution of the Earth's atmosphere

Table 12.1 Composition of Earth's current dry atmosphere. (Data from Sflps, 2017.)

Component	Symbol	Percent by volume	Mixing ratio	Concentration, PPM
Nitrogen	n_2	78.084%	0.78	780.840.0
Oxygen	o_2	20.945%	0.209	209.460.0
Argon	Ar	0.934	0.00934	9,340.0
Carbon dioxide	CO_2	0.036	0.00036	360.0
Neon	Ne	0.00182	0.0000182	18.2
Helium	He	0.000524	0.0000052	5.24
Methane	ch_4	0.00015	0.0000015	1.5
Krypton	Kr	0.000114	0.00000114	1.14
Hydrogen	H_2	0.00005	0.0000005	0.5

Table 12.2 Barometric pressure at various altitudes (elevations).

Altitude, ft.	Barometric pressure, mm. Hg.
O (sea level)	760
5000	632
8,000	570
10.000	529
20,000	350
40.000	140
60,000	50

composition and pressure, in the absence of nitrogen bacterial consumption, is presented in Figure 12.1.

Today's atmospheric pressure is shown in Table 12.2. The partial pressure of each gas component is defined, as the pressure that would be excreted by the molecules of gas × if all the other gases were removed from the mixture of atmospheric gases. Because many of the concentrations of the gases have a low partial pressure, one can use Dalton's ideal gas law, which states that the partial pressure (p_x) of gas × is related to the total pressure, p_t = atmosphere, by the mixing ratio, C_X:

$$p_X = C_X p_t , \qquad \text{(Eq. 12.1)}$$

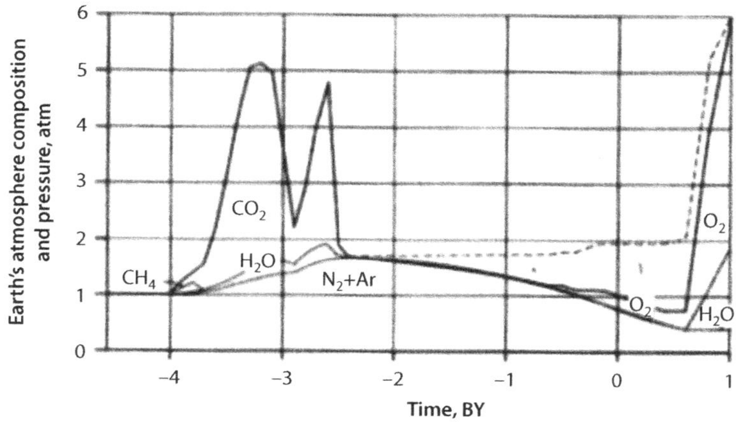

Figure 12.1 Earth's atmosphere composition and pressure evolution. The dashed line is the atmospheric pressure in the absence of bacterial consumption of nitrogen.

Or, the partial pressure (p_x) is equal to the total pressure times the mole fraction, y_x, of a particular gas in a mixture of gases:

$$p_x = p_t\, y_x \tag{Eq. 12.2}$$

The value of the total atmospheric pressure, p_t, is presented in Table 12.2 at various altitudes. The partial pressure of a gas is a measure of the frequency of collisions of gas molecules with surfaces or other molecules.

As noted in Figure 5.1, when the Earth was formed from the debris of a nebula, its atmosphere contained very low amounts of inert nitrogen and just traces of noble gases but higher quantities of hydrogen and helium. The Earth's atmospheric pressure at that time did not exceed 1 mbar. During the Archaean time, the primordial atmosphere contained none of the chemically active gases (e.g., CO_2, CO, O_2, or H_2O) as all these components were rapidly absorbed by the regolith covering the surface of the Earth. Due to the high surface temperatures, the hydrogen and helium molecules had an escape velocity adequate to escape the Earth's atmosphere.

The Earth's degassing of the mantle began during the Early Archaean time. The Earth's atmospheric pressure comprised the partial pressures of carbon dioxide, nitrogen, and methane. At that time, the carbon dioxide content in the atmosphere was increasing due to the CO_2 reduction of iron (see Figure 5.1). The atmosphere was substantially reducing and consisted of carbon dioxide, nitrogen, and methane. Beginning at about 3.5 BY ago, after a significant rise in the atmosphere's average temperature (see Chapter 14), the atmosphere's carbon dioxide-nitrogen composition was amended by the partial pressure of water vapor. After the methane dissociation (≈3.8 BY ago), the Earth's atmosphere became a neutral carbon dioxide-nitrogen gas-shell.

The nitrogen partial pressure noticeably increased in the Late Archaean time due to nitrogen degassing from the mantle. The carbon dioxide pressure began declining at the end of the Archaean time, because the atmospheric carbon dioxide was intensely being bonded within carbonates. After the core separation and the formation of the Earth's crustal serpentine layer (see Figure 5.15), which emerged because of the severe decline in tectonic activity, almost the entire carbon dioxide in the atmosphere was bonded in carbonate rocks.

During the Proterozoic time, the Earth's atmosphere became almost purely nitrogen with minute admixture of argon and methane (see Figure 12.1). Nitrogen's partial pressure may have reached 1.4 to 1.5 atm in the Early Proterozoic time. Beginning in the Middle Proterozoic time, the Earth's atmospheric pressure began to decline noticeably due to the activity of nitrogen-consuming bacteria. Simultaneously, in the Late Riphean, oxygen started to accumulate rapidly in the Earth's atmosphere, especially after a total disappearance of the metallic iron from the Precambrian mantle.

During the Phanerozoic time, nitrogen's partial pressure continued to decline, although in Paleozoic and Mesozoic it was amply compensated by the accelerated accumulation of biogenic oxygen. Upon evolution of the flowering plants, which were the major oxygen generators at the end Mesozoic, the Earth's atmosphere's pressure reached ≈230 mbar. Oxygen together with nitrogen continued to increase the atmospheric pressure until it reached ≈1 atm. Later, in the Cenozoic, Earth's atmospheric pressure began dropping again. This atmospheric pressure decline resulted in the cooling of the Earth's climate. This may have noticeably affected the activity of the nitrogen-consuming bacteria, which, in turn, resulted in a decreased nitrogen partial pressure.

Nitrogen-consuming bacteria have played a positive role in the evolution of the Earth's climate, creating favorable conditions for the evolution of life on Earth. Without the bacteria and the production of nitrogen, the current atmospheric pressure would be approximately 2 atm and the resulting average Earth's surface temperature would be over 180 °C at the equator. The latter is above the coagulation temperature of some vitally important proteins. It is likely that favorable conditions for life may have been present only on the mountaintops at high latitudes (lower temperatures). Vitally important oxygen could not have accumulated in sufficient amounts under such extreme conditions. In fact, if not for nitrogen removal from the atmosphere, the Earth would be presently (as in Archaean) populated only by the thermophilic bacteria and maybe the primitive multicellular organisms.

In the future, there will be a drastic oxygen partial pressure increase in 600 MY. This will be due to abiogenic oxygen degassing, as a result of the core separation process after the complete oxidation of the mantle iron to the stoichiometry. At that time the atmospheric pressure will rapidly rise and will exceed 10 atm due to the rise of oxygen content and partial pressure increase of the Earth's atmosphere.

Using the derived atmospheric composition and pressure, and the earlier proposed adiabatic theory of the greenhouse effect (Sorokhtin, 2001), it is possible to calculate the temperature parameters of the former, present, and future Earth's climates using Eq. 2.26.

Figure 12.1 was prepared, estimating the total atmospheric pressure and its varying content of components over geologic time. A sharp increase in the oxygen partial pressure in the future (≈600 MY), due to the increase in free abiogenic oxygen (gas not produced by the action of living organisms) from the Earth's mantle degassing process (Sorokhtin and Uskakov, 2002). This increase in free oxygen is the result of reducing the silicate iron oxide:

$$2FeO \rightarrow Fe \cdot FeO + O. \qquad \text{(Eq. 12.1)}$$

The released oxygen from this reaction, under high pressure and release of the compression energy (due to a smaller volume of the magnetite) again combines with FeO, forming the magnetite component of the mantle:

$$3FeO + O \rightarrow Fe_3O_4 + 76.48 \text{ kcal/mole}. \qquad \text{(Eq. 12.2)}$$

After a total oxidation of the mantle iron silicate to magnetite formation, of the Earth's core matter will have to be accompanied by the release of free oxygen:

$$2Fe_3O_4 \rightarrow 3Fe \cdot FeO + 5O. \qquad \text{(Eq. 12.3)}$$

As a result, in the future, additional free oxygen will enter the atmosphere, increasing the oxygen content as indicated in Figure 12.1. The partial pressure of oxygen will increase above 10 atm as a direct result of this additional oxygen influx. Eq. 2.26 indicates that this increase in pressure will raise the Earth's surface temperature above 180 °C. The Earth would be similar to the planet Venus. After evaporation of the ocean water, the atmospheric pressure will rise further and the Earth's temperature will exceed 600 °C. Surface temperatures of this magnitude will evaporate the oceans and destroy all life on the Earth. It should be noted that the current surface temperature on Venus is 460 °C, which is cooler than that of our future Earth, as it has a much higher carbon dioxide content in its atmosphere.

Part IV

VARIOUS FACTORS AFFECTING THE EVOLUTION OF THE EARTH'S CLIMATE

13

Earth's Orbital Distance from the Sun

Effect of Gravity on Earth's Orbital Paths

Every object in the universe is attracted to every other object in the universe by a force called gravity. Neutron identified mass and separation-distance as the important constituents of the universal force of gravity, *F.* Gravity, g, is one of the weakest forces in the universe. Between any two objects, the force, *F*, is dependent on the quantity of mass of each object and the inverse of the square of the distance between their centers of gravity.

$$F = \frac{gm_1m_2}{d^2} \tag{13.1}$$

where: g is the gravitational constant, $g = 6.674 \times 10^{-11}$ N (m^2/kg^2); m_1 (kg) is the mass of the first object; m_2 (kg) is the mass of the second object and d, is the distance (m) between the centers of mass of the two objects.

The gravitational force, *F*, between two objects can be very strong, even if only one of the objects is massive. If the objects are separated by a distance, *d*, the force quickly weakens as the distance, *d*, between the centers of gravity increases. It should also be noted in Eq 13.1, that d, because it is squared, reduces the force, F, between the two objects. An excellent example of this is the comparison of forces, is the difference of the influence between the Sun and the moon on the tides of the Earth's ocean. The Sun is the most massive object in our solar system, with a distance from the Earth of $d_{S-E} \approx 1.50 \times 10^8$ km. It is 333,000 times more massive than the Earth, exerting less force (F_{S-E}) on the Earth's ocean tides than the moon (F_{M-E}) at a distance $d_{M-E} \approx 3.825 \times 10^5$ km. Yet the mass of the moon is 1/6 the mass of the Earth. Another example of how distance is the most crucial factor in determining the gravitational force is that of a human standing on the Earth (F_{E-h}). Inasmuch as the distance between the Sun and a person on the surface of the Earth (d_{S-h}), is much greater than the distance between the Earth's center of gravity and the human (d_{E-h}) the Sun's gravitational force on the human (F_{S-h}) is far smaller than that which the Earth exerts on that person (F_{E-h})

The moon's force on a human is 0.06% of the Sun's force. The same reasoning demonstrates that the force a person feels from the moon is even smaller, or 0.00035% of the Earth's gravitational force. The force from the planet Jupiter, when it is closest to the Earth, is even lower, only 0.0000037% of the Earth's gravitational force (Univ. Calif. Santa Barb., 2016).

The gravitational force between the Sun and the Earth (F_{S-E}) is about 3.54×10^{22} N. This force is sufficient to keep the Earth orbiting around the Sun. Although this is the primary force that dictates the Earth's orbit about the Sun, gravitational forces from other planets, within our solar system, also affect the Earth's orbit about the Sun to a much smaller degree. The gravitational force of the moon on the Earth is ≈0.55% of the gravitational force between the Sun and the Earth. When Jupiter and Mars are closest to the Earth, Jupiter exerts 0.0062% of this force on the Earth's orbit and Mars only 0.00023% (UCSB, 2016). Other large bodies within the solar system directly affect the Earth's orbit about the Sun, to a much lesser degree, altering the distance (d_{S-E}) and thus the quantity of energy the Earth obtains from the Sun.

Earth's Orbital Path About the Sun

Newton recognized that the reason that one object orbits about another is related to the same gravitational force, F, that controls why objects fall to the Earth's surface when they are dropped. As the most massive object in the solar system, the Sun's gravity strongly pulls on every object within the solar system (Figure 13.1). All bodies within the solar system also exert

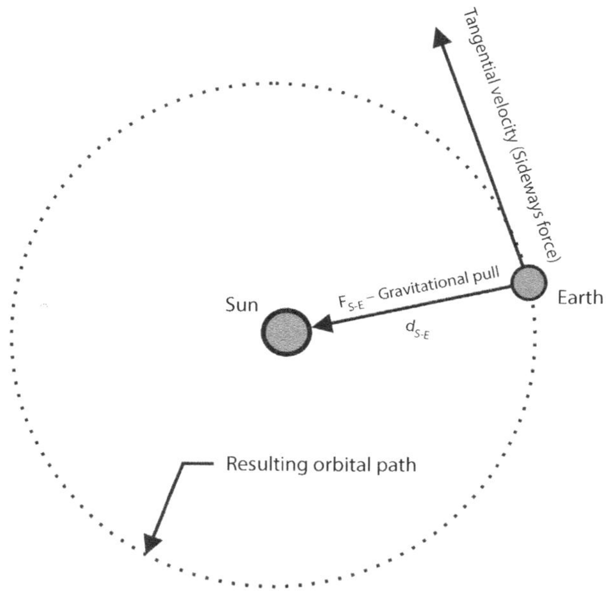

Figure 13.1 Schematic showing the forces controlling the orbit of the Earth about the Sun due to gravity.

a force (attraction) on each other; however, the force of the pull on the Earth's orbit, is significantly lower than that of the Sun's, as they have much less mass and, in many cases, often a greater distance, *d*, between their centers of gravity (between them and the Earth). Thus, because the Sun has a significantly greater mass than the other planets, all bodies (either directly or indirectly) orbit the Sun (solar system). When smaller solar objects are close enough to larger bodies, e.g., moons close to planets, the bodies with a larger mass are often capable of capturing these objects, forcing them to orbit about the larger body, while both bodies orbit the Sun. The Sun exerts the strongest gravitational pull of any object in our solar system and dominates the solar orbital system. Newton's law can be used to describe the total gravitational force:

$$F = ma, \qquad \text{(Eq. 13.2)}$$

where m represents the objects mass and a is the gravitational acceleration. The force, F in Eq. 13.1 may be used to compute the force between two objects; however, it should be realized that other objects may also have a significant influence on the total force excerted on the object. The total force and its direction is a summation of all the forces on the object.

After examining Figure 13.1, the following question might be asked: "*If the Sun is pulling the planet toward itself, why doesn't the planet just collide with the Sun*?" The answer is: "*in addition to falling toward the Sun, the planet also has a tangential velocity or movement sideways to the Sun within the elliptical plane, generating a force perpendicular to the force of the Sun's gravity*". When the force generated by this sideways velocity of the object and the *force pulling away* equals the gravity (*falling toward*) force, the object orbits the Sun (Figure 13.1). If the tangential velocity is greater, the force pulling away from the Sun is greater than the Sun's gravity force, then the object escapes the pull of the Sun's gravity and leaves the solar system. If the tangential velocity is less, the force pulling away is weaker than the Sun's pull of gravity, the object will crash into the Sun. This is the same force, *F*, one encounters when a weight is fastened to a string and is swung about oneself. As the weight swings around, the string exerts a pull away from one's hand, just as a planet attempts to pull away from the Sun; however, the force of gravity, represented by the pull on the string, prevents the weight escaping. Increasing the tangential velocity of the weight, increases the pull on the string, requiring more force to keep the weight in orbit. Without the tangential velocity, the planet would fall toward the Sun. Likewise, without the pull toward the Sun, the planet would go flying off in a straight line, escaping its orbit. This is what would happen if you let go of the string. For a planet to be held in orbit about the Sun, the force, F_{away}, of the object must be equal to the gravitational force, F_{toward}, pulling it toward the Sun. The F_{away} is dependent upon the tangential velocity of the planet. At higher velocities, the planet will move a greater distance, *d*, away from the Sun and at lower velocities, it will move closer to the Sun. Figure 13.1 is a schematic illustrating the forces holding a planet in orbit about the Sun. If the tangential velocity is either too high or too low, the planet can either escape or crash into the Sun. The tangential velocity to the Sun of a planet determines the orbital distance, *d*, of the planet from the Sun. The faster the velocity, the further the planet is from the Sun. In effect, a planet orbiting the Sun is continually falling toward the Sun. Planets that are further from the Sun must have greater tangential velocities to remain in orbit.

Kepler's Laws Pertaining to Planetary Orbits

For a two-body interaction (e.g., Sun and a planet) Kepler's three laws of planetary motion state that: (1) *Law of Orbits*: All planets move about the Sun in elliptical orbits with the Sun at one of the elliptical focus of the orbital path (Figure 13.2). (2) *Law of Areas*: A line that connects a planet to the Sun sweeps out equal areas for equal lengths of time (Figure 13.3); and

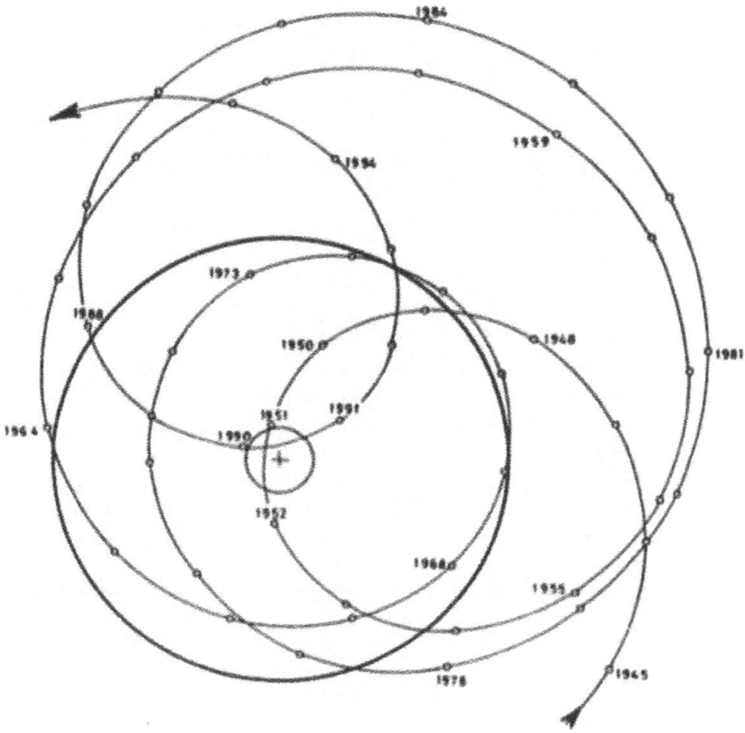

Figure 13.2 Sun's motion around the solar system's center of mass where the Sun's limb is shown as a thick solid circle. The motion of the Sun's center of mass between 1945 and 1995 is shown by the thinner curves with the years listed. (After Landscheidt, 2003.)

(3) *Law of Periods*: The square of the period for any planet is proportional to the cube of the semi-major axis of its orbit (Figure 13.4) (Daugherty, 2006).

At all times, as the planets travel along their orbit about the Sun, energy is exchanged and conserved. The momentum of the planet is always conserved as described by Newton's *Law of Conservation of Energy*:

$$\textit{Potential Energy} + \textit{Kinetic Energy} = \textit{Constant} \qquad \text{(Eq. 13.3)}$$

The potential energy, *PE* is:

$$PE = mad\,, \qquad \text{(Eq. 13.4)}$$

where m represents the mass of the planet and d the distance between the Sun and the planet. a is the acceleration of the mass of the planet caused

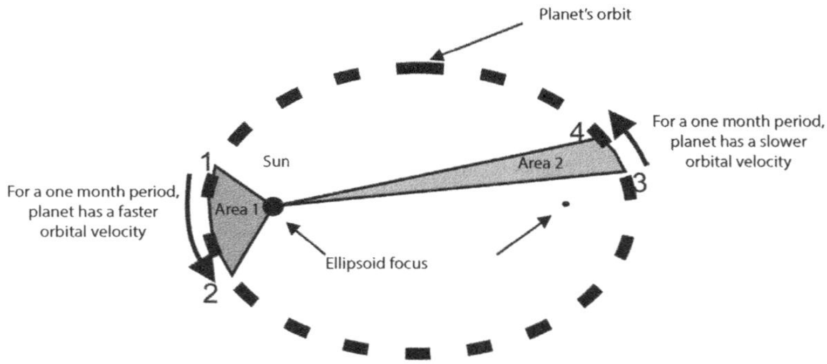

Figure 13.3 Kepler's three laws of planetary motion are reflected in this schematic of the orbit of a planet about the Sun. *Area 1* is always equal to the area swept by *Area 2*. The planet's position on the orbital path requires that the gravitation forces are greater on the planet at the shorter distances from the Sun, *d*, resulting in greater tangential velocities along the path. Thus, the tangential velocity of the planet increases or decreases in velocity as it moves through its orbital path. At all times PE + KE = a constant.

by the force of gravitational attraction between the Sun and the planet. The kinetic energy, *KE* is equal to:

$$KE = \frac{1}{2}v^2, \qquad \text{(Eq. 13.5)}$$

where v is the velocity of the planet about the Sun. Newton's law of the conservation of energy describes why the planet varies its velocity as it follows its orbit about the Sun. Figure 13.2 is a schematic illustrating the mechanics of a planet orbiting about the Sun. Kepler recognized that all planetary orbits within our solar system are ellipses, with the Sun located at one focus (Figure 13.2). The distance, *d*, between the centers of gravity of the two bodies varies as the planet moves about its orbit (13.5). When the planet is the furthest, d_{max}, from the Sun, it is at its aphelion (apogee), the *PE* is at its greatest value (the distance, *d*, is at its greatest value) and the *KE* is the lowest value (tangential velocity is at its lowest). When the planet is closest to the Sun, d_{min}, it is at its perihelion (perigee) of its elliptical orbit, the *PE* is at its lowest value (the distance, *d*, from the Sun is at its shortest) and the *KE* is at its greatest value (the tangential velocity is at its maximum value). Because *d* is the shortest distance from the Sun at its the planet at its perigee, the quantity of energy transferred from the Sun to the planet is also the greatest, resulting in a higher global temperature for that portion of its orbit. The planet's axis angle of tilt determines the quantity (concentration) of energy transferred to the planet (see discussion in Chapter 3).

Two factors control the quantity of energy transferred between the two bodies for a distance, *d*: (1) the quantity of energy transmitted at that distance and (2) the difference in the tilt between the two bodies or the angle of precession controls the concentration of energy transmitted (for a more detailed discussion, see Chapter 3).

Figure 13.6 describes the quantity of energy that is transmitted over a distance, *d*. Mathematically this quantity of energy, *e*, transmitted over a distance, *d*, decreases by a factor of:

$$\text{Energy transmitted, } e = e_o\left(\frac{1}{d^2}\right), \qquad \text{(Eq. 13.6)}$$

where e_o is the initial quantity of energy being transmitted at $d = 0$.

Eccentricity of an Object's Orbit

The measure of the difference in shape between a circle and ellipse is referred to as eccentricity, ε. The eccentricity for a circle is zero and the value ε varies from 0 to 1 as the orbit becomes more elliptical, tending toward a straight line for ε = 1. The orbital eccentricity of the solar system's planets is listed in Table 13.1. This table shows that planets, other than Pluto which has a large eccentricity (ε = 0.248), have almost circular orbits. However, since Pluto is no longer considered as a planet, Mercury now has that honor, with ε = 0.2056. The eccentricity of an orbit may be calculated by the following equation:

$$\varepsilon = \left(\frac{d_{max} - d_{min}}{d_{max} + d_{min}}\right), \qquad \text{(Eq. 13.7)}$$

where ε is defined as the coefficient of variation of the maximum and minimum orbital axis.

As indicated in Figure 13.3, a line that connects a planet in its orbit to the Sun, for an equal period of time, sweeps out equal areas. In this example, movement from point 1 to 2 (area A) is a much longer orbital distance than that from point 3 to 4 (area B). In this case, the same time is required for the planet to sweep both areas A and B. The distance of the orbital path from point 1 to 2 is shorter than from point 3 to 4, but because the tangential velocities between the two points differ, the time elapsed for sweeping both areas is the same.

Table 13.1 Planetary orbit dimensions. (Data from Unv. Calif. Santa Barb., 2016.)

Planet	Orbital, semimajor axis (astronomical Units)	Orbital period, years	Orbital eccentricity, e
Mercury	0.387	0.241	0.206
Venus	0.723	0.615	0.007
Earth	1.000	1.000	0.017
Mars	1.524	1.881	0.093
Jupiter	5.203	11.86	0.048
Saturn	9.539	29.46	0.056
Uranus	19.19	84.01	0.046
Neptune	30.05	164.8	0.010
Pluto	39.53	248.6	0.248

Kepler also recognized the *Law of Periods*, noting that the square of a planet's period, T, is proportional to the cube of the semi-major axis, a, of the planets orbit (Figure 13.4):

$$T^2 = \frac{4\pi^2}{GM}a^3, \qquad \text{(Eq. 13.8)}$$

where $T =$ time (yr); $a =$ astronomical units (AU); and $M =$ solar mass.

Table 13.2 is the NASA's planetary fact sheet for our solar system. It should be noted that the mass of the Sun at 1.989×10^{30} kg is so dominant in the solar system, that: (1) all bodies in the solar system revolve about the Sun; (2) the planetary orbits are almost circular, ($\varepsilon \approx 0$) and, (3) although the other planets exert a gravitational force on each other, the effect they have on each other's orbital path is significantly small.

The following equation can be utilized for calculation of orbits of moons about planets where the mass of one body does not overwhelm the mass of the orbiting body:

$$T^2 = \frac{4\pi^2}{G(M_1 + M_2)}a^3 \qquad \text{(Eq. 13.9)}$$

Effect of Other Planets on Earth's Orbit

The Sun is positioned at one elliptical focus where the solar system orbits about it. Unlike the Sun with most of the solar system's mass, the other

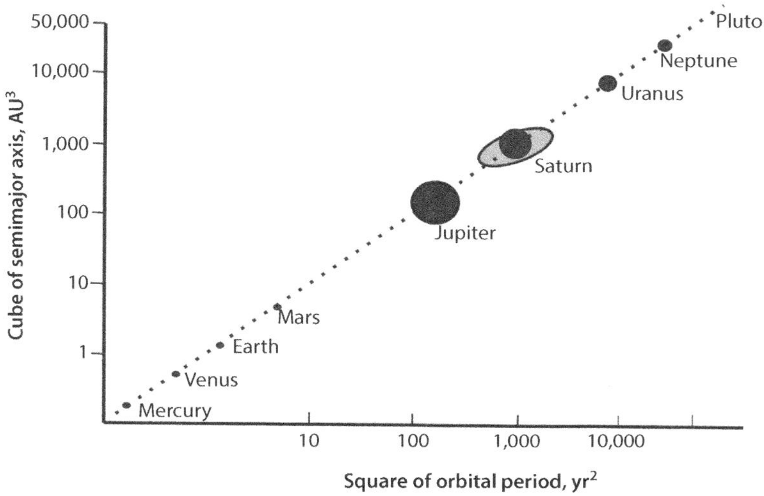

Figure 13.4 Schematic illustrating Kepler's Law of Periods. The straight dotted line expresses Kepler's Law of Periods. (Data from Daugherty, 2006.)

planets have only a minor gravitational effect on Earth's orbital path due to their small mass (see Table 13.2). The Sun contains about 99% of the mass of the solar system and forces all bodies within the solar system to orbit about it. However, other planets do exert a gravitational force on: (1) the orbital path of the other planets and (2) those bodies (moons) which orbit the planet while also revolving about the Sun. All bodies within the solar system exert a gravitational force on each other's orbital path within the solar system.

Table 13.2 lists several massive planets that have sufficient mass to affect the Earth's orbital path. Jupiter is the second most massive object within the solar system and far more massive than the other planets. The effect of Jupiter's gravitational pull on the Earth's orbital path varies with the distance (d_{E-J}) between the centers of gravity of the two planets. The overall gravitational force of Jupiter on Earth's orbit is much smaller than that of the Sun due to: (1) the greater distance (d_{E-J}) of the Earth to Jupiter and (2) the lesser mass of Jupiter compared to that of the Sun. Likewise, even the smaller massive planets, e.g., Saturn, Neptune, etc., affect the Earth's orbital path, but to a far lesser degree as they are at a greater distance and have significantly less mass. Thus, all the material in the solar system to some degree influence the final shape of the Earth's orbit. The distance (d) between the Earth and Sun directly affects the quantity, or concentration, of the energy transmitted (see Figure 13.6), reaching the Earth's surface and determining the Earth's temperatures.

Table 13.2 Planetary fact sheet in United States units. (Data obtained from NASA, 2018; https://w.w.w.nssdc.gsfc.nasa.gov/planetary/-planetfact.html.)

	Mercury	Venus	Earth	Moon	Mars	Jupiter	Saturn	Uranus	Neptune	Pluto
Mass (1021 tons)	0.364	5.37	6.58	0.81	0.708	2093	627	0.5.7	113	0.0161
Diameter (miles)	3032	7521	7926	2159	4221	88,846	74,897	31,763	30,725	1464
Density (lbs3.39/ft^3)	339	327	344	209	246	83	43	79	102	131
Gravity (ft/s^2)	12.1	29.1	32.1	5.1	12.1	75.9	29.4	28.5	36.0	2.3
Escape Velocity. (miles/s)	2.7	6.4	7.0	1.5	3.1	37.0	22.1	13.2	14.6	0.8
Rotation Period (hrs)	1407.6	−5832.5	23.9	655.7	24.6	9.9	10.7	−17.2	16.1	−153.3
Length of day (hrs)	4222.6	2802.0	24.0	708.7	24.7	9.9	10.7	17.2	16.1	153.3
Distance from Sun (10^6 miles)	36.0	67.2	93.0	0.239*	141.6	3670.0	890.8	1784.8	2793.1	3670.0
Perihelion (10^6 km)	28.6	66.8	91.4	0.226*	128.4	460.1	840.4	1703.4	2761.6	2756.9
Aphelion (10^6 km)	43.4	67.7	94.5	0.252*	154.9	507.4	941.1	1866.4	2824.5	4583.2
Orbital Period (days)	88.0	224.7	365.2	27.3	687.0	4331	10,747	30,589	59,800	90,560
Orbital Velocity (miles/s)	29.4	21.8	18.5	0.64	15.0	8.1	6.0	4.2	3.4	2.9
Orbital inclination (deg.)	7.0	3.4	0.0	5.1	1.9	1.3	2.5	0.8	1.8	17.2
Orbital Eccentricity	0.205	0.007	0.017	0.055	0.094	0.049	0.057	0.046	0.011	0.244
Obliquity to Orbit (deg.)	0.034	177.4	23.4	6.7	25.2	3.1	26.7	97.8	28.3	122.5
Temperature, mean (°F)	333	867	59	−4	−85	−166	−220	−320	−330	−375
Surface pressure (atm.)	0	91	1	0	0.01	unknown	unknown	unknown	unknown	0.0001
Number of Moons	0	0	1	0	2	67	62	27	14	5
Ring System	No	No	No	No	No	Yes	Yes	Yes	Yes	No
Global Magnetic Fields	Yes	No	Yes	No	No	Yes	Yes	Yes	Yes	Unknown

*Relative to Earth, not Sun

The patterns of interaction between the solar system's Sun and its revolving bodies, due to gravitational forces, started as early as the initial accretion stage from the protoplanetary cloud. The Sun ejects a high-energy stream of charged particles, often referred to as the solar wind, that sweeps all gas and volatile components from the Sun into the distant periphery of the solar system. This is a continuous flow of charged particles streaming from the Sun in all directions at speeds of ≈400 km/s or ≈1 million miles per hour. The source of the solar wind is the Sun's hot temperature of the corona. The temperature of the corona is so high that the particles within the corona have velocities that are at and/or above the escape velocity from the Sun's gravity. The solar wind is not uniform, although it is always directed away from the Sun. It changes speed and carries with it (1) magnetic clouds, (2) interacting (turbulent) regions where higher speed wind catches up and mixes with slower speed winds, and (3) has compositional variations. The chemical differentiation of the protoplanetary cloud's matter was affected by the heating of the Sun on the disk's main area, at the earlier stage of its compression and especially after nuclear reactions were initiated within it (Sorokhtin, 2006).

Because of the mechanisms of interaction, condensation occurred in the central areas of the protoplanetary disk, mostly refractory elements and compounds with high ionization potential of refractory metals, e.g., Fe, Ni and oxides of Al_2O_3, CaO, MgO, Ti_2O_3, SiO_2, Cr_2O_3, FeO. At the same time, there were concentrations of easily fusible and ionized elements (e.g., Li, Na, K, Rb). Ba and rare earth elements (e.g., Hg, & Pb) within this part of the protoplanetary cloud were very low. To a lesser degree, the elements forming the Earth were impoverished in sulphur, zinc, tin and some other elements. The gas components, e.g., H_2, He and other noble gases, H_2O, CO, CO_2, CH_4, NH_3, H_2S, SO_2 and SO_3, HCl, HF were almost completely swept out of the internal areas of the protoplanetary cloud and concentrated at its periphery, where they subsequently formed the giant-gas-planets possessing massive and high-density gas shells, e.g., Jupiter (see Table 13.2) (Sorokhtin *et al.*, 2013).

Jupiter is the second most massive object in the solar system, and it has the second strongest effect, although weaker, on the shape of Earth's orbital path. The overall gravitational force of Jupiter on Earth's orbital path is much smaller than that of the sun due to Jupiter's far greater distance and lesser mass than the Sun. Likewise, the smaller, less massive planets, e.g., Saturn, Neptune, etc., affect Earth's orbit to a slighter degree than Jupiter. At any time along Earth's orbit, the distance of the Sun to the Earth directly affects the quantity of energy reaching the Earth's surface (see Figure 13.6).

Although the gravitational influence of Jupiter and Saturn, along with the lesser bodies of the solar system, affect the Earth's final orbital path

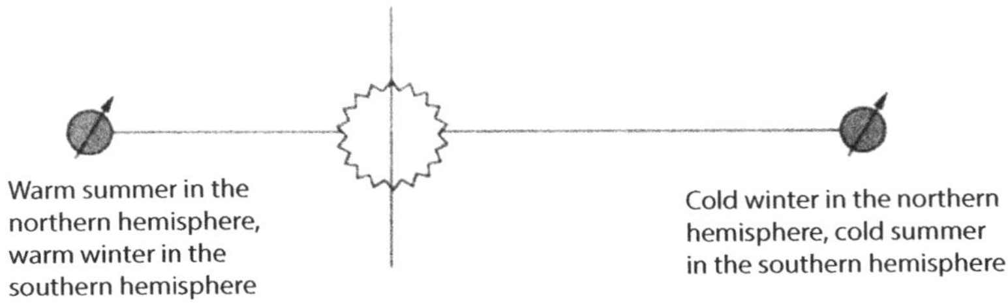

Figure 13.5 The Earth's yearly hemispheric contrast in distance from the Sun, d, between the summer and winter. The Sun is located in one of the ellipses foci, in Earth's elliptical orbit of revolution about the Sun. The arrows point to the North star and indicate the angle of the Earth's rotation on its axis. The vertical line shows the center of the Sun's mass.

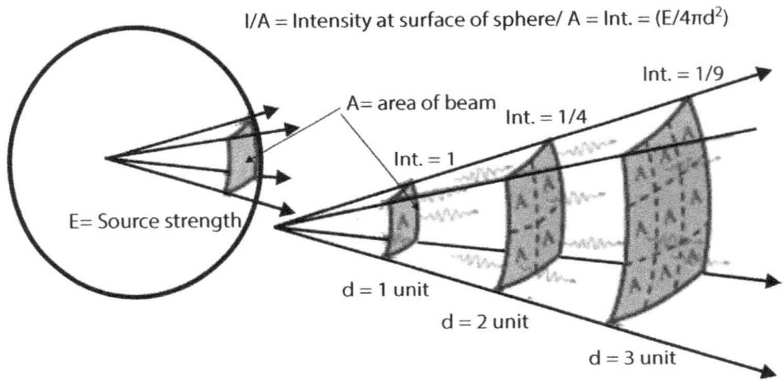

Figure 13.6 Schematic showing spread of energy, intensity/area of ray, with distance from the source.

about the Sun, they also affect the position of the Sun and other bodies within the solar system (Figure 13.2). The result is a complex picture of Earth revolving around the Sun in an elliptic orbit, with average orbital ellipticity of ≈0.017 (Table 13.2). Thus, the Earth during a single revolution (i.e., in one year), varies with the distance, *d*, from the Sun (which is in one of the foci of Earth's elliptical orbit); the change in the other planet positions with time directly affects Earth's orbital path (Figure 13.5). No two sequential years have the same orbital path, resulting in a cyclical variation of temperature with respect to time.

Today, the ellipticity for the Earth's orbit and average distance of the Earth to Sun is ≈149 million km. The pericenter of Earth's orbit is ≈1.48 million km (92 million miles) and the apocenter, ≈1.53 million km (95.1 million miles) (see Figure 13.7). Therefore, Earth to the Sun distance,

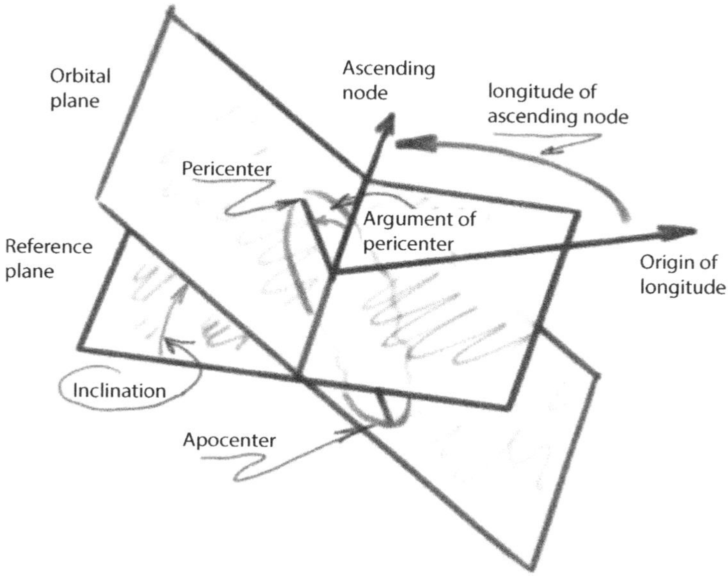

Figure 13.7 Parameters describing elliptical orbits: (1) Apocenter – The location of the greatest distance between the orbiting body and the central body when the orbit is an ellipse. The apocenter is diametrically opposite the pericenter on the major axis of the orbit. (2) Pericenter—The shortest distance between the center of the orbiting body and the center of the orbited body. (After Cornell, 2017. https://www.classe.cornell.edu/~seb/celestia/orbital-parameters.html.)

d, varies by ≈5 million km (3.1 million miles). The solar insolation of Earth is always in inverse proportion with the squared distance between the Sun and Earth. This results in the annual variation of Earth's temperature, which is affected by the Sun's constant between 1.41 to 1.32×10^6 erg/cm² or an average $S_0 = 1.367 \times 10^6$ erg/cm² · s. According to Monin and Sonechkin (2005), the Sun to Earth distance, *d*, varies between 1.47- to 1.52-million km (91.4 and 94.5 million miles) with the Sun constant, between 1.428- to 1.322×10^6 erg/cm². The following equation:

$$T = b^{\alpha}\left[\frac{S(1-A)}{\sigma\left(4\left(\frac{\left(\frac{\pi}{2}-\psi\right)}{\frac{\pi}{2}}\right)+4\left(\frac{\psi}{\frac{\pi}{2}}\right)\left(\frac{(1-\cos\psi)}{(\sin\psi)^2}\right)\right)}\right]^{\frac{1}{4}}\left(\frac{p}{p_o}\right)^{\alpha} \quad \text{(Eq. 2.25)}$$

shows that variations of the solar constant may correspond to the average values of the near-surface temperature of 288.5 K and 290.2 K, with fluctuations of up to 4.7 °C. These fluctuations, however, have a seasonal nature and apparently have negligible effect on the Earth's average temperature.

As a result, when the Earth is coming through the perihelion, its average temperature annually increases by 2.2 °C and in aphelion, on the contrary, declines by 2.5 °C. Depending on the Earth's revolution axis orientation relative to the direction to the Sun, climatic contrasts in different seasons appear (see Figure 13.5). At the onset of a cooler winter and warmer summer in the Northern Hemisphere and simultaneously a warmer winter and a cooler summer in the Southern Hemisphere. Examples of such contrasting seasons are the severe winters and hot summers of 1940-1941 or on the contrary, warm winters and cool summers of 1983-1984. It should be kept in mind that the real picture of climatic variations is much more complex due to superposition of the Sun's motion around the solar system mass center and the Earth's revolution around the Sun.

The Effect of the Planet Jupiter on Earth's Orbital Path

Jupiter is the largest planet in the Solar system and its mass, $m_J \approx 1.90 \times 10^{24}$ kg is ≈318 times greater than that of the Earth, $m_E \approx 5.97 \times 10^{24}$ kg. Minimum distance between the Earth and Jupiter ($d_{J\text{-}E\text{-}min}$) (at the planet's opposition point on the apsis) is approximately 6.3045×10^{13} cm, where the maximum distance ($d_{J\text{-}E\text{-}max}$) (also on the apsis) is $\approx 9.3045 \times 10^{13}$ cm. Therefore, the maximum gravitational pull from Jupiter on the Earth is equal to:

$$F_{J-E_{max}} = \gamma \left[\frac{m_J \times m_E}{d_{J-E_{min}}} \right] = 1.936 \times 10^{23} \text{ dynes}, \qquad \text{(Eq. 13.10)}$$

and the minimum is:

$$F_{J-E_{min}} = \gamma \left[\frac{m_J \times m_E}{(d_{J-E_{max}})^2} \right] = 8.758 \times 10^{22} \text{ dynes}. \qquad \text{(Eq. 13.11)}$$

where $\gamma = 6.67 \times 10^{-8}$ cm^3/g-s^2 is the gravitational acceleration.

The gravitational pull on the Earth by the Sun is greater by an order of magnitude than that of the Earth by Jupiter. The mass of the Sun,

$m_S \approx 1.99 \times 10^{33}$ g, and the distance between the Sun and the Earth is much smaller: $d_{S-E} \approx 1.496 \times 10^{13}$ cm. In this case the average gravity pull by the Sun on the Earth is equal to:

$$F_{S-E} = \gamma \left[\frac{m_E \times m_S}{(d_{SE})^2} \right] = 3.5166 \times 10^{27} \text{ dynes}. \qquad \text{(Eq. 13.12)}$$

The difference between the gravity forces by the Sun and Jupiter on the Earth's orbital path at the moment of their crossing the apsis is to:

$$\Delta F_{\min} = 3.5166 \times 10^{27} - 1.936 \times 10^{23} = 3.5164 \times 10^{27} \text{ dynes}, \qquad \text{(Eq. 13.13)}$$

$$\Delta F_{\max} = 3.5166 \times 10^{27} - 8.758 \times 10^{23} = 3.5165 \times 10^{27} \text{ dynes}. \qquad \text{(Eq. 13.14)}$$

Thus, the difference between the maximum and minimum forces acting between the Jupiter and the Earth at their opposition and that of Jupiter in the contrary position is:

$$\Delta F = 3.5165 \times 10^{27} - 3.5164 \times 10^{27} \approx 1 \times 10^{23} \text{ dynes}. \qquad \text{(Eq. 13.15)}$$

Other planets within the solar system, e.g., Saturn and Mars, render a much smaller effect on Earth's orbital path. The interaction forces between Saturn and Earth at the point of planetary opposition is $F_{Sat\text{-}E} = 1.395 \times 10^{22}$ and for Mars and Earth is $F_{Mar-E} = 4.605 \times 10^{21}$ dynes.

Earth's revolution around the Sun occurs around the center of mass between the Sun and Jupiter. This center, with the average Jupiter-Sun distance of $d_{S-J} \approx 7.8 \times 10^{13}$ cm, is offset from the Sun's center of mass by:

$$x = d_{S-J} \left[\frac{M_J}{M_S} \right] = 7.454 \times 10^{10} \text{ cm}. \qquad \text{(Eq. 13.16)}$$

This distance is outside of the Sun's body, as the radius of the Sun's is $r_S = 6.96 \times 10^{10}$ cm.

The time required for one revolution of Jupiter around the Sun is 11.862 years. The period of interaction between the Earth and Jupiter, and of Jupiter's effect on the position of Earth's orbit at any time would also be about 11.862 years. One might anticipate that under the pull of gravity from the external planets, the Earth's orbit would be changing its direction of ellipticity over the same period. Earth's revolution around the Sun represents a huge *gyroscope* and thus is quite stable (see Table 13.3). The apsis running through the center of the Sun and connecting the points of perihelion (closest approach of Earth to the Sun) and aphelion (largest distance between the Earth and the Sun) is quite stable. This precession of the Earth equals 93,000 years and over the foreseeable future, one would not anticipate the position of the apsis to change appreciably. The effect of this cycle on Earth's temperature is reflected in its 100,000-year cycles as reflected on the long-term temperature charts (see Figure 10.17).

The minimum and maximum distances between the Sun and the Earth can be calculated by Eq. 13.16:

$$d_{S-E_{\min}} = 1.5\times10^{13} - 7.457\times10^{10} = 1.4925\times10^{13}\,\text{cm},$$

and

$$d_{S-E_{\max}} = 1.5\times10^{13} + 7.457\times10^{10} = 1.5075\times10^{13}\,\text{cm}.$$

The average value of the solar constant $S = 1.367 \times 10^6$ erg/ cm$^2 \cdot$ s. Thus, the maximum and minimum values of the solar constant are:

$$S_{\max} = 1.367\times10^6\left[\frac{1.5\times10^{13}}{1.4925\times10^{25}}\right]^2 = 1.381\times10^6\,\text{erg/cm}^2\cdot\text{s},$$

and

$$S_{\min} = 1.367\times10^6\left[\frac{1.5\times10^{13}}{1.5075\times10^{25}}\right]^2 = 1.353\times10^6\,\text{erg/cm}^2\cdot\text{s}.$$

At the Earth's precession angle, $\omega = 23.44°$, albedo $A = 0.3$, normal atmosphere of pressure, $p = 1$ atm., and temperature in degrees Kelvin is:

$$T = 1.093\left[\frac{S(1-A)}{3.5\sigma}\right]^{\frac{1}{4}}, \qquad \text{(Eq. 13.17)}$$

where $\sigma = 5.67 \times 10^{-5}$ erg/cm$^2 \cdot$ s $\cdot$ deg^4 (the Stefan-Boltzmann constant). Thus:

$$T_{\max} = 1.093\left[\frac{(1.381\times10^{6})0.7}{\left(5.67\times10^{-5}\right)3.5}\right]^{\frac{1}{4}} = 288.756\ \mathrm{K},$$

and

$$T_{\min} = 1.093\left[\frac{(1.353\times10^{6})0.7}{(5.67\times10^{-5})3.5}\right]^{\frac{1}{4}} = 287.281\ \mathrm{K}.$$

The difference between these temperatures is $\Delta T \approx 1.475$ °C. This difference in value is caused by the Jupiter gravitational effect. Although small, it could likely increase due to the synoptic activity of the atmosphere and changes in the Earth's albedo, e.g., due to the snow cover in winter with simultaneous increase in the cloud cover. The value of the factor k in the following equation, for this increase is unclear:

$$k = \frac{F_S - F_Y - F_{St} - F_M}{F_S + F_Y + F_{St} + F_M} = 0.999876. \qquad \text{(Eq. 13.18)}$$

In the authors' estimate, it may be much greater than 2 as the annual average temperatures practically do not change.

The elongation of the Earth's elliptical orbital path is in proportion to the difference in the gravity pull from the Sun and the closest external planets (Mars, Jupiter and Saturn). The largest eccentricity of Earth's orbit occurs at the point of its aphelion during the planetary alignment, i.e., when the positions of Mars, Jupiter and Saturn coincide. The maximum eccentricity is determined by the following equation:

$$\varepsilon^2 = (1 - k^2). \qquad \text{(Eq. 13.19)}$$

Thus, $\varepsilon \approx 0.01578$. This eccentricity value is close to the observation data: $\varepsilon = 0.0167$. The small difference, most likely, is due to inaccuracies in the determination of planetary masses and their distances from the Earth. However, the largest effect on the Earth's orbital path is exerted primarily by Jupiter:

$$k_J = \frac{F_S - F_J}{F_S + F_J} 0.999886 \text{ and } \varepsilon_J = 0.01507, \qquad \text{(Eq. 13.20)}$$

which is responsible for 95.5% of the Earth's eccentricity.

The question is, how does Jupiter's gravity affect the Earth's elliptical orbit? When the Earth's apsis corresponds with the direction toward Jupiter, some offset of the focus (in which the Sun is positioned) of the Earth's elliptical orbit occurs toward the center of the Sun-Jupiter mass by the amount determined by the interrelation of forces acting from the side of Jupiter and the Sun:

$$\Delta d_{S-J} = d_{S-J}\left[\frac{m_J}{m_S}\right] = 7.8045\times10^{13}\left[\frac{1.9007\times10^{30}}{1.99\times10^{33}}\right] = 7.4543\times10^{3}\,\text{cm}.$$

(Eq. 13.21)

A similar offset of the focus length occurs both at Jupiter's intersection of apsis from the side of aphelion and of the perihelion. For this reason, the distance between the Earth and the Sun, affected by Jupiter's gravity pull, changes from: $d_{S-E\min} = (1.5\times10^{13}) - (7.4543\times10^{10}) = 1.4925\times10^{13}\,\text{cm}$, (Eq. 13.22) to

$$L_{S+E\max} = (1.5\times10^{13}) + (7.4543\times10^{10}) = 1.5075\times10^{13}\,\text{cm}. \quad \text{(Eq. 13.23)}$$

According to the adiabatic theory of the greenhouse effect (Sorokhtin 2006; Sorokhtin *et al.*, 2007, 2011), the average absolute temperature of the planet's troposphere T (in degrees Kelvin) is a function of the Sun constant, S, and the atmospheric pressure, p, as reflected in Eq. 2.22, where $S = 1.367 \times 10^6$ erg/cm^2 · s is the solar constant (flow of the solar energy reaching the Earth); $\sigma = 5.67 \times 10^{-5}$ erg/cm^2 · s · °C^4 is the Stefan-Boltzmann constant; A is the planet's reflectivity (albedo), for the Earth $A \approx 0.3$; p_o is the unit of pressure, e.g., $p_o = 1$ atm; the scaling factor, $b = 1.597$; a is the adiabatic exponent, $\alpha = (\gamma - 1/\gamma)$; $\gamma = (c_p/c_v)$ where c_p and c_v are the specific heats of gas at constant pressure and constant volume, respectively; ψ is the angle of precession (Earth's value today is $\psi = 23.44°$). At $\psi = 23.44°$, the denominator in Eq. 3-27 is equal to 3.502 rather than 4.0 in the classic format at $\psi = 0$.

Changes in the solar constant by 0.1 to 0.2 % in Eq. 2-26 result in changes of Earth's near-surface temperature by 0.072 to 0.143 °C. Much larger temperature fluctuations are caused by the Earth's axis precession. Considering the quoted Earth's orbit eccentricity value and the average of the Earth to the Sun distance ($d_{S-E} \approx$ 150 MM km), it is possible to determine that the Earth's orbit perihelion is ≈147.4 MM km and its aphelion ≈152.6 MM km. This shows that the change in Earth to the Sun distance may exceed 5 MM km. The solar insolation of Earth is always inversely proportional to the square of the Sun-to-Earth distance, d_{S-E}, causing annual fluctuations in the solar constant by 1.416 to 1.321 × 10^6 erg/cm^2 · s, with an average value of $S_0 = 1.367 \times 10^6$ erg/cm^2 · s. These changes in the solar constant value, shown in Eq. 2.26, correspond with the Earth's average near-surface temperature of 285.5 to 290.5 K, or its fluctuations by up to ±2.5 °C. These fluctuations, however, are seasonal and apparently have no significant effect on the Earth's average annual temperature.

Thus, during the Earth's passage of the perihelion, its average temperature approximately increases annually by 2.5 °C; in the aphelion, on the other hand, it decreases by about the same 2.5 °C. Depending on the Earth's revolution axis direction (relative to the direction to the Sun), this results in climatic contrasts during the opposite seasons. For instance, the onset of a cool winter and hot summer in the Northern Hemisphere and simultaneously warm winter and cool summer in the Southern Hemisphere (warm autumn and cool spring or warm spring and cool autumn) at the average climatic parameters of these seasons (see Figure 13.5).

Earth's climatic fluctuations are likely much more complex due to the superposition of the Sun's motion around the solar system's center of mass, the movement of the other planets effecting the Earth's orbital path around the Sun. As shown in Figure 10.17, these fluctuations in Earth's orbital path are affected by: (1) the gravity pull from the other planets, primarily Jupiter and Saturn, and (2) the Sun delineates rather complicated trajectories in its movement around the solar system's center of mass as shown in Figure 13.2. Together with the Sun, the Earth's orbit delineates similar *pirouettes*. The Earth's elliptic orbital focus is always close to the Sun's center of mass, thus the general image of the Earth revolving around the Sun on the elliptic orbits is complex.

The pull of gravity from the planets revolving around the Sun, affect not only the orbit of the Earth and all other planets, but also the position of the Sun itself relative to the solar system's center of mass and their orbital paths (see Figure 13.2).

Temperature fluctuations of ±2.5 °C caused by the Earth's revolution around the Sun on an elliptic orbit, influence the amplitude of the Earth's seasonal temperature changes. One period of cyclical influence by Jupiter on the climate is ≈11.86 Earth years, whereas the Earth's seasonal temperature fluctuations are exactly 12 months. Thus, the superposition of the large planets' gravitational effect (mostly Jupiter's) on Earth's climate should result in a modulation of the seasonal temperature fluctuations with the amplitude of ±2.5 °C and a total cyclical period of ≈12 years. An estimate of such climatic disturbances of seasonal temperatures at the Earth's

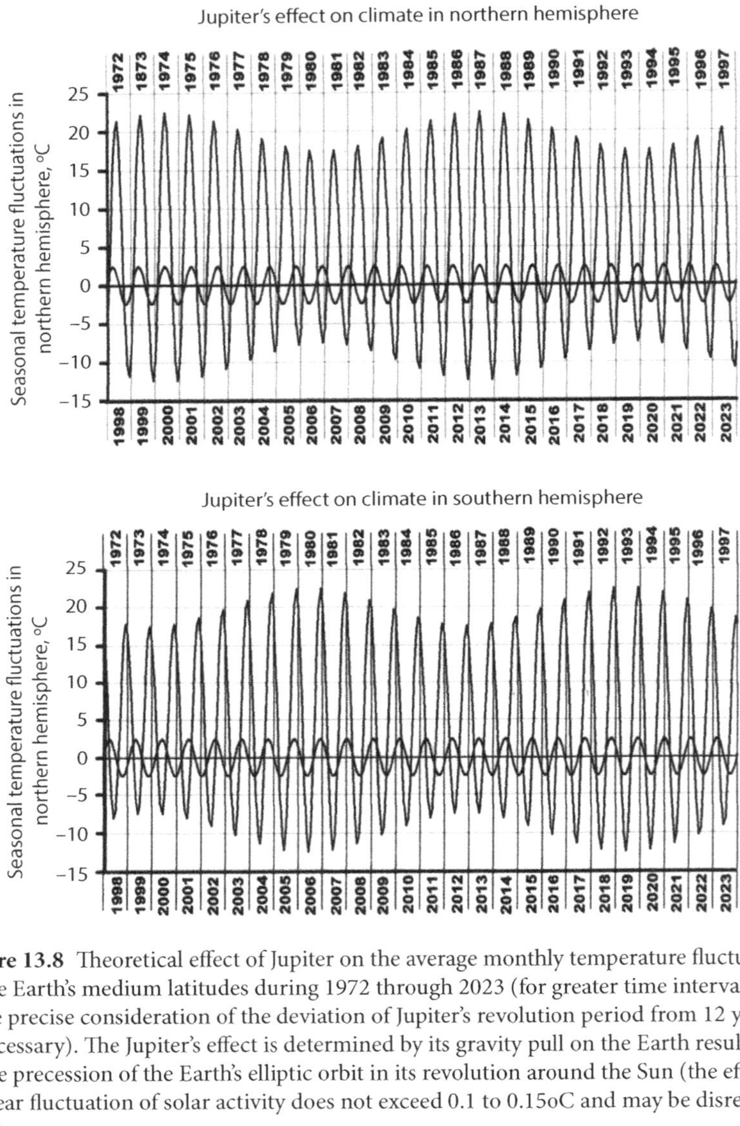

Figure 13.8 Theoretical effect of Jupiter on the average monthly temperature fluctuations in the Earth's medium latitudes during 1972 through 2023 (for greater time intervals, a more precise consideration of the deviation of Jupiter's revolution period from 12 years is necessary). The Jupiter's effect is determined by its gravity pull on the Earth resulting in the precession of the Earth's elliptic orbit in its revolution around the Sun (the effect of 11-year fluctuation of solar activity does not exceed 0.1 to 0.15oC and may be disregarded here).

mid-latitudes is included in Figure 13.8. The quoted climatic disturbances are usually outside the range of vision of scientists, as the very effect of 13-year modulation by Jupiter's influence disappears due to averaging of seasonal fluctuations in the processing of observations.

Figure 13.8 displays four-time intervals (1978–1983, 1991–1996, 2004–2009 and 2017–2022) in the Northern Hemisphere, which exhibited abnormally warm winters and cool summer seasons. On the other hand, the periods 1972–1978, 1985–1990, 1997–2002 and 2010–2015 had cold winters and warm summers. In the Southern Hemisphere, the situation is inversed. Time intervals of 1978–1983, 1991–1996, 2004–2009 and 2017–2022 had cold winters and warm summer seasons, whereas the periods 1972–1978, 1985–1990, 1997–2002 and 2010–2015 had warm winters and cool summers. This regularity, of course, is quite understandable as cold seasons occur primarily when the Earth goes through the aphelion (i.e., the point of largest distance from the Sun). Exactly in a half a year, when the Earth arrives at the perihelion (i.e., closest to the Sun), the Earth warms up. For instance, in 2006 Earth went through the perihelion on December 3rd and the aphelion on June 6th of 2007. That is why the winter of 2007 in the Northern Hemisphere turned out to be unusually warm and the summer relatively cool, whereas in the Southern Hemisphere the winter was cold and the summer warm.

In 1815, the summer in Europe was cold. The origin of this phenomenon has been attributed to the Tambora volcanic eruption on the Sumbava Island, Indonesia. But contrary to all expectations, the winter was warm and not cold. Again, an analogous situation appears to have occurred in the Northern Hemisphere in 2007: a relatively cool summer (except for rare days when the warm air from the arid belt of the Mediterranean area arrived at the medium latitudes of the Northern Hemisphere) and a very warm winter. If one subtracts 1815 from 2007 we get 192, which is exactly 16 cycles of Jupiter's gravitational action on the Earth's climate ($12 \times 16 = 192$).

In 2006, the temperatures in fall were warmer, December was relatively warm and the snow, as in 1826, fell only in January. The interval between these two events was exactly 180 years (2006-1826). This time interval approximately constitutes 15 cycles of Jupiter's effect on Earth's climate: $12 \times 15 = 180$ years (another option is $11.862 \times 15 = 177.93 \approx 178$ years).

In 1940 and 1941, Europe experienced exceptionally severe winters with bitter frosts in December. Again, a similar climatic situation was observed in 2000 and 2001. The time difference between these events was 60 years or five 12-year cycles of Jupiter's gravitational effect on the Earth's climate. Subtracting from the year 2000, the 60-year period (which is equal to five 12-year cycles) gives the year 1940. Comparable results are obtained for the year 2001: ($2001 - 12 \times 5$ or 1941).

A very warm winter occurred in European Russia in 1947 when at New Year's Eve one could get his feet wet. Again, there is a 12×5 or 60-year cycle between 2007 and 1947.

This climatic data is approximate and does not consider the effect of continent and ocean positions, ocean currents, and the synoptic activity on the real temperature distribution occurring above the Earth's surface. This may be the reason why comparison of theoretical climatic temperature regime changes with empiric monthly average temperature distributions for the Northern and Southern hemispheres (Figure 13.8) often does not reveal the anticipated data correlation.

The likely reasons for the total mismatch of temperatures may be the result of a strong negative feedback (through the planet's albedo) between the incidental solar radiation and the near-surface temperature. Indeed, any increase in the solar insolation of the Earth's temperature will result in an increased water evaporation from the oceans and, thus, in an increased cloud cover and the Earth's reflectivity (albedo) and decreased solar energy reaching the Earth's surface. As a result, the surface temperature declines to a new equilibrium level. Under a strong negative feedback, such temperature decline may be very substantial, drastically decreasing fluctuations at the set albedo. Thus, the temperature changes due to fluctuations of solar radiation become negligible on the background of stronger synoptic temperature fluctuations, which apparently does occur.

Despite the weakened solar radiation effect on changes in the near-surface temperatures, Jupiter influences the energy processes in the Earth's atmosphere through humidity condensation (the water heat capacity ≈ 1 cal/g$\cdot$deg is significant). It is possible, therefore, that the periodicity in climatic anomalies (droughts, rainy seasons and especially storms) is associated with fluctuations in the solar insolation. This includes those occurring due to the Earth's revolving around the Sun in elliptical orbits, i.e., eventually under the effect of Jupiter's gravity pull on the Earth. The stated climatic changes must also be affected by its complicating long-period fluctuations of the solar insolation associated with the Sun's revolution around the center of mass of the solar system (see Figure 13.2) and by the prolonged changes in the solar activity (most likely, by nuclear synthesis processes within the Sun).

Currently we live near the maximum of a temporary climate warming, which began as early as the seventeenth century when there was no human effect on the climate by the *greenhouse gases* released into the atmosphere. This shows that the current warming is clearly of a natural origin and soon will unavoidably be replaced by a new cooling period. Reflecting on the temperature charts in Chapter one, although the Earth is experiencing temperature cycles, the overall temperature of the Earth has been steadily decreasing over the past 65 MY.

A similar conclusion was also reached by the well-known German climatologist Landscheidt (2003). He published an article: *New Little Ice Age Instead of Global Warming?* In his estimate, regardless of the recent continuing increase of carbon dioxide in the atmosphere by man's activities, there will be only cooling and the onset of a *little ice age* comparable to the one experienced in the seventeenth and eighteenth centuries. The anticipated average temperature decline will be 1 °C or greater.

14

Climatalogical Effect of Continental Drift

Continental Drift's Effect on the Earth's Precession Angle

Continental drift is an explanation of how the Earth's continental crust shifts its position over time. This concept set forth in 1912 by Alfred Wegener, a geophysicist and meteorologist, was an attempt to explain why similar mineral, animal and plant fossils are found on different continents with separations of oceans, mountains, etc. Wegener proposed that all the continents were once joined together in a gigantic continent, called an *Urkontinent,* before breaking up into large bodies of land and drifting to their current positions (Figure 14.1). At the time of his announcement, geologists denounced Wegener's theory of continental drift. He published the details of his ideas in 1915, in a book entitled *The Origin of Continents and Oceans.* Today, this concept, that continents can and do shift position on the surface of the Earth, known as the plate tectonics, is widely accepted.

This shifting of the continent's position on the Earth's surface alters the Earth's spherical symmetry and center-of-mass. As a result of these

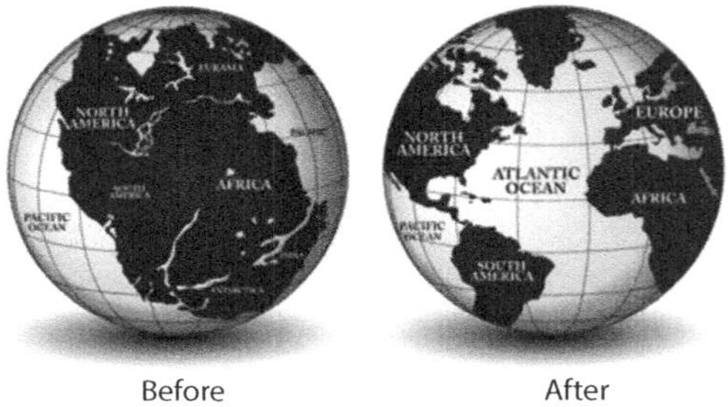

Figure 14.1 Earth before and after the land masses (continents) were broken and redistributed over the Earth's surface. (After Live Science, 2015.)

changes in continental position, there is a change of the Earth's precession angle. As discussed in Chapter 2, a change in the precession angle directly affects the planet's temperature regime as shown by Eq. 2.26:

$$T_t = b^{\alpha}\left[\frac{S(1-A)}{\sigma\left\{4\left(\frac{\frac{\pi}{2}-\psi}{\frac{\pi}{2}}\right)+2\left(\frac{\psi}{\frac{\pi}{2}}\right)\left(\frac{2}{1+\cos\psi}\right)\right\}}\right]^{\frac{1}{4}}\left(\frac{p_t}{p_o}\right)^{\alpha}. \quad \text{(Eq. 2.25)}$$

The influence of this change can be expressed mathematically, even though one may not fully recognize all the topographical density anomalies in the mantle and the Earth's crust. Today, the effect upon the precession angle, due to the current continental positions on the surface of the Earth, is relatively small, due to their current distribution, the moment of force, $\Delta M_{cont+m} \approx 0.2$ to 0.6×10^{29} cm$^2 \cdot$ g/s^2. However, at the time of the formation of the supercontinents, the positioning of the land mass was often located at lower latitudes and distributed in only one area on the Earth's surface. At the time of the supercontinents, the Earth's center of mass would have been different than today's, as would the moment of force ΔM_{cont+m}. Figure 14.2 estimates many of the recognized continental positions on Earth's surface over the past 2.6 BY.

Chumakov (2004) determined surface temperatures of the Sargasso Sea from ocean sediments. ≈100-MY ago in the tropical belt, approximately at 20° North latitude where temperatures ranged between 25° to 30 °C. The maximum equatorial temperature, T_{eq}, likely reached 28 to 32 °C, whereas

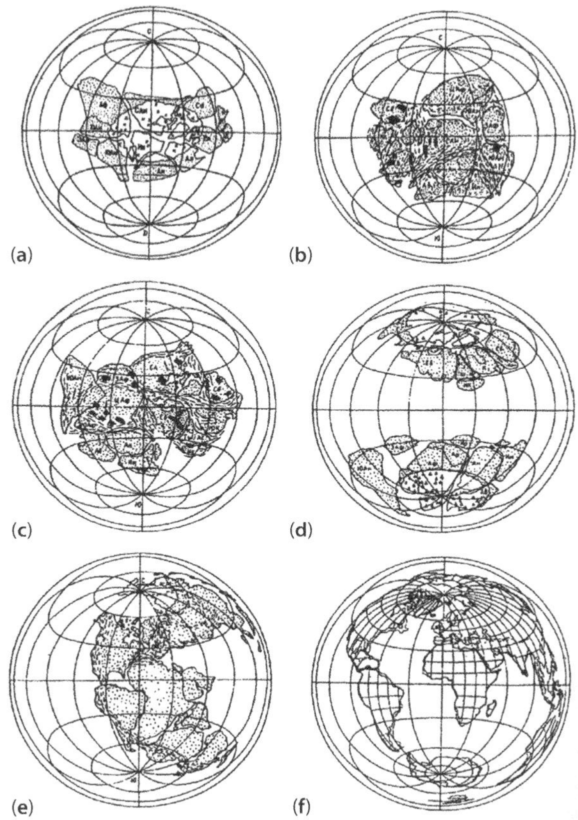

Figure 14.2 Paleo-reconstruction of the positions of the continents and oceans over the past 2.6 BY according to Lambert: (a) Monogaea, 2.6 BY ago. White areas on the continents indicate glaciers, whereas crosses indicate tillites and tilloids (Chumakov, 1978). (b) Megagaea by Sttille, 1.8 BY ago. Wavy shading indicates folding belts and blackened areas show Redbed Formations (Anatolieva, 1978). (c) Mezogaea, 1 BY ago. (d) Disintegration of Mezogaea and formations of Laurasia and Gondwana, 750 MY ago. White areas on the continents indicate glaciers, whereas small triangles show location of tillites and tillotds (Chumakov, 1978). (e) Pangaea of Wegener, 200 MY ago, and (f) Present-day position of continents and oceans. (After Sorokhtin *et al.*, 2009, figure 1.41, p. 79.)

the Earth's poles were dominated by positive temperatures, T_{pl}. Recognizing the temperature relationships:

$$T_{eq} = T_s + 0.036\ \Delta T \text{ and } T_{pl} = T_s + 0.64\ \Delta T\,, \qquad \text{(Eq. 14.1)}$$

and

$$\Delta T \approx \frac{\Delta T_o}{p_s} \qquad \text{(Eq. 14.2)}$$

where the present-day difference between the temperatures at the equator and poles of the Earth, ΔT_o, is about 30 °C. Also, regarding the moment of force (ΔM_m):

$$\Delta M_m + M_s = \Delta M_{cont+m}. \qquad \text{(Eq. 14.3)}$$

One can estimate the equatorial temperature at ($T_{eq} \approx 32$ °C) and the positive polar temperatures (T_{pl}) are compatible with a precession angle of $\psi = 34°$. Beside the Cretaceous paleotemperature, determination for the precession angle in the past geologic epochs, this calculation is supported by data from the glaciation-distribution era in the Earth's history.

The stationary precession angle at the time of the Pangaea formation on the surface of the Earth, ≈100 MY ago, approached 34°. Judging from the reconstruction of other super continents and their different configurations with smaller individual masses (see Figure 14.2), the estimates of precession angles at the time of their formation (Figure 14.3) are somewhat lower (Sorokhtin and Ushakov, 2002). During the time intervals between the

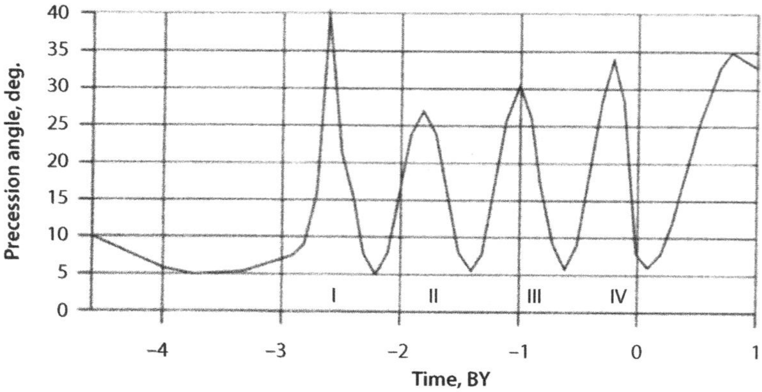

Figure 14.3 Probable values of the precession angle of the Earth at the time of the supercontinents: (I) Monogaea, (II) Megagaea, (III) Mezogaea (Rodinia); and (IV) Pangaea. The tremendous masses of the supercontinents of Monogaea, Megagaea, and Mezogaea were 0.7, 0.8, and 0.9 of the mass of Pangaea, respectively. It was also assumed that a future supercontinent Hypergaea would form in 800 MY due to the Earth's weakening tectonic activity.

formation of the supercontinents, the writers estimated that the precession angle was close to a minimum of 6° to 8° C.

Using the Earth's precession angle in (Figures 14.3 and 14.4), one can estimate the Earth's average surface temperature. In addition, one can use the same model and acknowledging the changes in surface temperatures for the oceans and continents, at Earth's equator and poles, an estimate of the Earth's surface temperature may be determined.

During periods of supercontinent formation, as shown in Figures 14.3 and 14.4, were accompanied by a corresponding increase in the Earth's near-surface temperature. For example, during the formation of the supercontinent, Monogenea, ≈2.6 BY ago, the Earth's average temperature exceeded 70 °C at sea level. Such elevated temperatures during the Archean time were largely determined by the presence of a dense atmosphere and increasing precession angle during formation of a second supercontinent, ≈1,8 BY ago, the surface temperature of the oceans and continents were about 32 °C and 22 °C, respectively. The Earth's temperature had decreased to 26 °C, whereas the average temperature of the continents increased to 25 °C, due to the lowering continental height. Finally, during the formation of the supercontinent Pangaea, ≈200 MY ago, the temperature was 23 °C and lower. During the intervals between the formation of the supercontinents, the average temperatures at sea level and continents decreased by 7° to 10 °C.

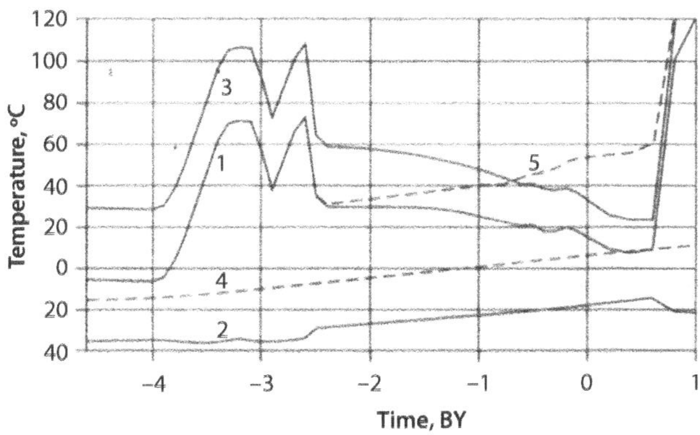

Figure 14.4 Evolution of Earth's temperature (climates) assuming a constant Earth's precession angle (current) of y = 24°. *Curve 1* – an average for Earth's surface temperature at sea level calculated using Eq. 2-25. *Curve 2* – the effective Earth's temperature using Eq. 2.23. *Curve 3* – Earth's atmospheric greenhouse effect using Eq. 2-6. *Curve 4* – The absolute black body temperature using Eq. 2.3. using the distance of the Earth-from-the-Sun, (d) describing the increase with time of the Sun's luminosity, *S*. *Curve 5* – Average Earth's temperature on the assumption that there was no nitrogen consumption by bacteria.

There was a drastic increase in the Earth's precession angle at the time of formation of the supercontinent Monogaea. This supercontinent formation was simultaneously associated with descending mantle flow, while on the opposite side of the equatorial zone, the matter from the former primordial Earth's kernel was ascending, being displaced by the Earth's core emergence (Sorokhtin, 2004; Sorokhtin and Ushakov, 2002). This ascendance of matter from depth would have created a substantial "bulge" of the mantle, substantially distorting the Earth's spherical symmetry and altering the Earth's center of gravity and, thus, its precession angle.

The formation of supercontinents results in changes other than just the physical structure. Because of the alteration of the location for the center of mass, it also alters the Earth's precession angle and, thus, the climatic conditions of the Earth. When these major climate changes occur, they can often be correlated with the beginning or closure of a geologic period. The correlation of the precession angle with the geologic event for supercontinent formation can be seen in the physical changes of the direction and intensity of oceanic currents, gyres, etc. Through geologic history, there have been several major climatic changes that can be attributed to changes in the precession angle. Today, the physical blockage of the equatorial oceans current system appears to coincide with icehouse events (Gerhard *et al.*, 2001). ***The conclusion is that pronounced global climatic changes are caused primarily by changes in insolation and those Earth's tectonics that affect the precession angle.*** Changes in insolation are often the result of a change in the precession angle between the Earth and Sun.

Latitudinal Temperature Contrast on Earth's Surface

To determine the latitudinal contrast for the Earth's climate, one can compare the planet Earth to Venus. Table 2.2 was prepared utilizing the known temperature measurements of Venus's atmosphere for low (0° to 30°) and high (75°) latitudes for different elevations, i.e., at different pressures. These Venus observations indicate that climatic contrast, as defined by temperature difference, for different latitudes is approximately proportional with the atmospheric pressure. Utilizing this relationship, it is possible to calculate the latitudinal variations of the temperature distribution within the Earth's atmosphere, dependent on its pressure. This enables the determination of not only the average Earth's temperature, but also the temperature at its poles and equator. Utilizing this correlation and the Earth's precession angle variation model, one can estimate the average temperature of the World's Ocean, and the temperature at the Earth's equator and poles (see Figure 14.4 and Eq. 2.26).

During the Phanerozoic time, because of the increased oxygen partial pressure in the atmosphere, the phyto- and zooplankton productivity increased. The phyto- and zooplankton subsequently served as a source for hydrocarbon generation. Sea plankton productivity likely increased even greater during the warm Mesozoic time interval, especially in Cretaceous. This process was certainly facilitated by a vast Late Cretaceous sea transgression that flooded about a third of the continental lowland areas. At the same time the deposition of organic matter and the deposition rate, significantly increased in the newly formed and heated epicontinental seas. In the open ocean, the deposition rate drastically declined. It is possible that a well-known Mesozoic (especially Cretaceous) burst of the oil generation was associated with this warming climate and the extensive marine transgression over the dry land. The nature of this Late Cretaceous transgression and its influence on the oceanic biota has been reviewed in greater detail by Sorokhtin and Ushakov (2002).

The equation for the determination of the temperature evolution on ancient continents is:

$$\text{Grad } T = \frac{g}{c_p} \quad \text{(Eq. 14.4)}$$

This equation enables the calculation of the average tropospheric temperatures at various elevations as a function of the atmospheric heat absorbing capacity, $c_{p,}$ and gravity acceleration g. It provides the opportunity to determine the atmospheric temperature at any elevation up to $p \approx 0.2$ atm. The atmospheric heat conductivity may be determined by:

$$\alpha = \frac{\mathrm{R}}{\mu(c_p + C_w + C_r)}, \quad \text{(Eq. 14.5)}$$

and

$$c_p = \frac{\left[\left\{p_{N_2} c_{p(N_2)}\right\} + \left\{p_{O_2} c_{p(O_2)}\right\} + \left\{p_{CO_2} c_{p(CO_2)}\right\} + \left\{p_{Ar} c_{p(Ar)}\right\}\right]}{p}. \quad \text{(Eq. 14.6)}$$

Sorokhtin and Ushakov (2002) used these equations to estimate the continental standing height above sea level (see Figure 14.4). The average temperature fluctuations at Earth's equator and poles can be estimated using Eqs. 2.26 and 14.3 (see Figure 14.6).

Each supercontinent formation epoch was accompanied by increasing near-surface temperatures (see Figures 14.4 & 14.5). At the time of the supercontinent Monogaea's emergence, ≈2.6 BY ago, the average temperature at the ocean level exceeded 70 °C, whereas on continents it only reached 30 °C. The elevated temperature at the end of Archaean time was substantially determined by: (1) a dense layer of atmosphere; (2) an increase in the Earth's precession angle during the formation of the Earth's core; and (3) the formation of the supercontinent Monogaea. During the formation of the supercontinent Megagaea (Stille's) (≈1,8 BY ago) the average oceanic and continental temperatures approached 22° and 32 °C, repectively. During the formation of the supercontinent of Mezogaea (Rodinia) (≈1 BY ago), the average oceanic temperature declined to 28 °C, whereas average temperature on the continents, due to their lower standing level increased to 25 °C. During the formation of Wegener's Pangaea supercontinent (≈200 MY ago) the temperatures declined, respectively, to 23 °C and 19 °C. In-between supercontinent formation periods, the average temperatures at sea level and on the continents declined by 7° to 10 °C.

At the beginning of the oldest glaciation era (Archean, ≈3.9 to 3.7 BY ago), a young sea basin and embryonic future continental shield emerged and was positioned within a relatively narrow equatorial belt. As a result,

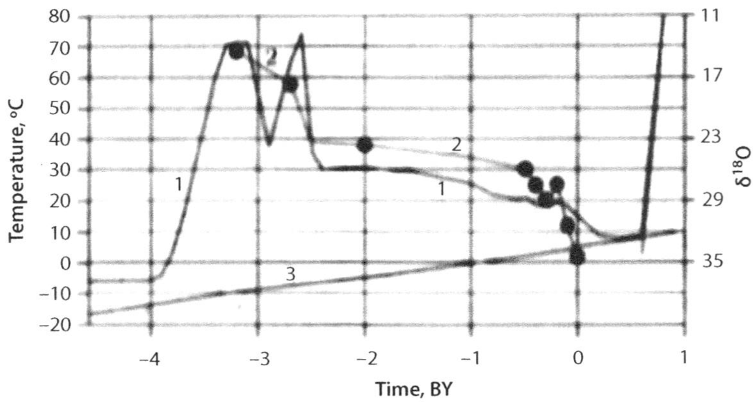

Figure 14.5 Isotopic temperature for ocean flints superposed on a theoretical curve of Earth's climate evolution. *Curve 1* – average theoretical curve of the Earth's temperature evolution at the oceanic surface (see Figure 14.5, curve 1); *Curve 2* – Oxygen $\delta^{18}O$ isotope shifts found in sea flints (see Figure 1.3, solid black dots, after Schopf, 1982) and estimated temperature of the depth of benthic water where the sea flints were deposited; *Curve 3* – Temperature of an absolutely black body at the Earth-Sun distance describing Sun luminosity (see Figure 1.10).

the sea basins of that time were warm, with temperatures ranging from (3° to 5 °C) to 20 °C (see Figure 14.4). The altitude of these young continental mountains may have reached an altitude of 3 to 4 km (Figure 14.5). Being on the equator and despite the general warming of the temperature of the Earth, the mountain temperatures were likely negative. The evidence of glaciations which did emerge in these elevated formations is local in nature; however, these glaciers are limited because it appears that there was very little liquid water present on the Earth. The authors suggest this may be an explanation of why there are no traces of early Archaean glaciation.

The Middle Archaean glaciation occurred, ≈2.9 BY ago (Figure 14.6), as the Earth's tectonic activity and the atmospheric pressure decreased (see Figures 4.4 and 5.3). The glaciation might have been developing in high and moderate latitudes, as the Earth's tectonic activity had broadened to latitudes in the range of ±40° to 50°. The young continental shields at the latitudes could have also drifted. At the same time, their standing height increased to 5.5 to 6 km (Figure 14.5). In the Middle Archaean time, the amount of water in the ocean basins significantly increased (see Figure 5.12). The temperature increased by 30° to 40 °C, increasing the evaporation. Therefore, the high-mountain glaciation in the Middle Archaean time most likely could have been accompanied by the formation of small ice sheets, which have left their footprint in the Earth's geologic record.

Sorokhtin (2004) noted that genuine oceans, although not very deep (only up to 1 km), emerged in the Early Proterozoic time. The continental stand remained rather high, ranging from 2 to 4 or 5 km (Figure 14.5).

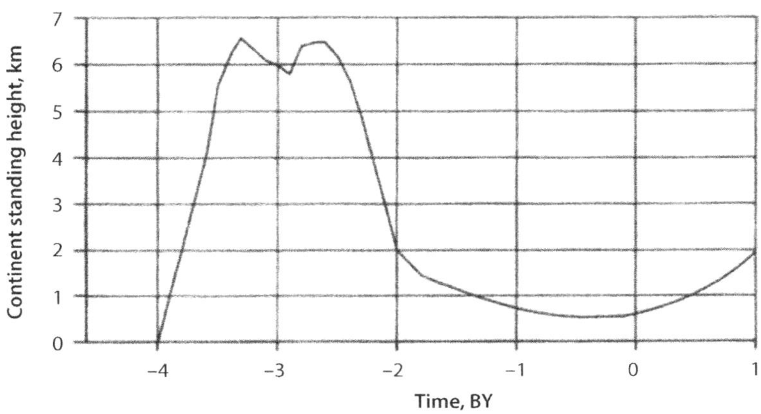

Figure 14.6 The standing height of continents versus time.

In the beginning of the third identified glaciation era (Huron glaciation, group III in Figure 14.6) glaciation sheets developed in all latitudes.

By the end of the third glaciation era, the first in the Earth's history, supercontinent Monogaea disintegrated. Its fragments, which had stood high above the sea level, had been scattered over all the Earth's surface. After this disintegration of the supercontinent, in the middle of the third glaciation era, 2.4 to 2.1 BY ago, the ice sheets probably formed on all continental massifs regardless of their latitude. Glacial deposits of this age testify to the nature of glaciations now found on today's continents. This includes several locations: in North America, the North American Huron formation; in Europe, the Baltic Shield; in South Africa, formations of Transvaal and Griqualand; and in Western Australia, Turi formation. In each of these regions, the glaciation age is almost identical, 2.4 to 2.2 BY ago.

During the time interval of 2 to 1 BY ago as shown in Figure 14.6, negative average annual temperatures did not exist on the Earth. Therefore, continental glaciation should not have occurred. Chumakov (2004) showed that there were no traces of glaciation existing in the horizons of Lower Proterozoic, Early Riphean, and most of the Middle Riphean. This was an extended glacier-free period in the Earth's evolution.

In the Late Riphean and Paleozoic, a new era of continental glaciation began. This period was associated with a notable cooling of the Earth's climate (see group IV in Figure 14.6). Vast snow-covered territories at the poles with a high albedo (around 0.7) resulted in substantially lower temperatures.

Despite rather severe continental glaciations, there were no stable marine glaciations (similar to those of the present-day, e.g., Arctic Ocean) either in the Late Riphean or in Paleozoic. That was because the circumpolar areas of the oceans and seas, at that time, had only a slightly negative average annual temperature. For that reason, melting occurred during the interglacial periods (see Figure 14.4).

The last and fifth glaciation period began in the middle of Cenozoic (≈40 MY ago), which was the most extensive in the Earth's history. The fifth glaciation era continues today. This is the last one in the Earth's history (group V, Figure 14.3). The first sea glaciations in the Earth's history occurred in the Oligocene and Miocene, in the Arctic Ocean and in portions of the Southern Ocean (shelf glacier covers in the Ross and Weddell seas). As a result, there was a substantial temperature decline of the near-bottom ocean water (16 °C to today's 2 °C) between the Paleocene and Eocene period.

Climate cooling existed during the Pleistocene on those continents positioned close to the poles. This resulted in negative average annual

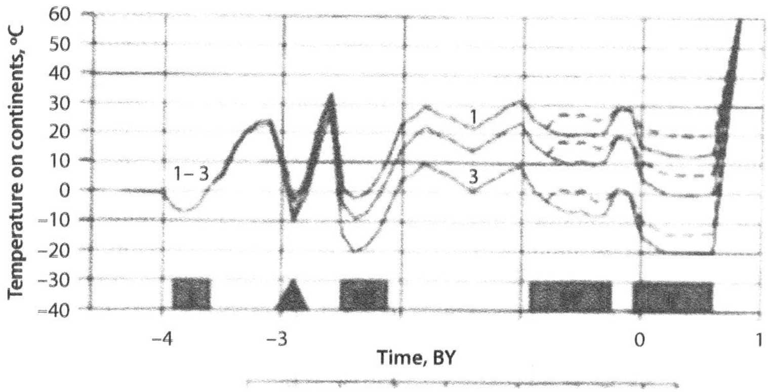

Figure 14.7 Evolution of the Earth's surface temperature for the continents caused by the mutual influence of: *Curve 1* – evolution of Earth's atmosphere; *Curve 2* – standing height of continents (see Figure 14.6); and Curve 3 – variation in the angle of precession (Figure 14.3). Temperatures calculated at: *Curve 1* – temperature at the Earth's equator of the continents; *Curve 2* – average temperature of continents; and *Curve 3* – temperature at the continental poles. Dashed line indicates temperatures during the interglacial periods I to VI: Periods of glaciation on high mountains. (After Chumakov, 2004.)

temperatures. As a result, the ice-sheets had to emerge on the continents. If the areal extent of these ice sheets had been significant, the albedo of the ice-sheet snow cover rendered an impact. Depending on the elevation of the glacial cover, the average temperatures above them dropped to –40 °C and at times, even –60 °C.

Verification of the theoretical estimates for the average temperatures on the surface of the ice sheets were obtained by comparing the estimates to the experimental data. Wells were drilled to obtain ice cores to determine the annual average temperatures $\overline{T}$ in the central areas of the Antarctic where possible the sea cyclones do not reach. Kotlyavov (2000) measured temperatures at Dome-C, ≈3.24 km above the sea level, $T = -53$ °C, and the Komsomolskaya and Vostok station, elevation 3.5 km above sea level, at a temperature, $\overline{T} = -55$ °C and –55.5 °C. O. G. Sorokhtin and A.P. Kapitsa measured a temperature of $\overline{T} = -60$ °C on the summit of the main Antarctic dome, at an elevation of 4 km above sea level. The theoretical temperature values for the same elevations are, respectively, –54.6°, –55.7° and –62.3 °C, which are close to the measured data.

The question arises as to why there was a warm climatic interval between two severe glacial periods at the end of the Mesozoic: between the Riphean-Paleozoic and the late Cenozoic era. The origin of this warm period was associated with two factors: (1) the formation of the supercontinent Pangaea, and (2) increased generation of biogenic oxygen.

The climate evolution theory expounded here indicates a substantial role of living organisms in the formation of the Earth's climates. The nitrogen-consuming bacteria have especially affected the global Earth's climate in the Late Proterozoic and Phanerozoic time. Their life-sustaining activity has more than halved the nitrogen partial pressure in the Earth's atmosphere. This lowering of the pressure has resulted in a significant cooling and emergence of glaciations in high and moderate latitudes of the continental Proterozoic and Phanerozoic. Generation of biogenic oxygen by phytoplankton and land vegetation during the Phanerozoic time gave rise to the blossoming of highly organized life on the Earth and partly compensated for the decline in partial pressure of nitrogen. That was the reason for a notable climate warming period occurred at the end of Mesozoic. However, after reaching by oxygen the maximum (equilibrium) pressure level of about 0.231 atm, a new atmospheric pressure decline and, thus, a new cooling era, with the succession of glacial epochs had begun.

Thus, the primary causes of the Earth's climate change have been a: (1) gradual decline in the atmospheric pressure due to life-sustaining activity of nitrogen-consuming bacteria; (2) variations in the precession angle; and (3) variations of the Sun activity.

15

Earth's Future Climate

Understanding the past allows one to better understand the future. As demonstrated earlier, one can use the historic Earth's atmosphere composition and pressure (Figure 5.3) as well as the adiabatic theory (Eq. 2.25) for the determination of the temperature parameters for the past, current and future Earth's temperatures. The calculated values of temperature from these equations (Eq. 2.25) are presented in Figure 15.1.

Figure 15.1 displays data illustrating that after the Archaean time, the Earth's surface temperature has been continuously declining despite the increase of the Sun's luminosity (*Curve 4*). After the accumulation of water on the Earth's surface, forming the ocean and emergence in the Early Proterozoic time of the oceanic crust, Earth's temperature decline has occurred due to bonding of carbon dioxide in carbonate rocks. Later, during the Proterozoic and Phanerozoic times, this process continued through the life-sustaining activity of nitrogen-consuming bacteria, which resulted in cooling down of the Earth's surface.

The effect of the nitrogen-consuming bacteria and thunderstorm activity on the Earth's climatic evolution played a positive role in creating a favorable environment for the evolution of life on the Earth. Had these

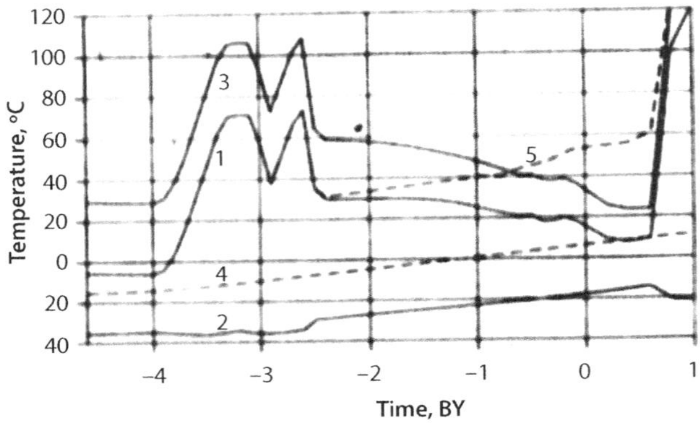

Figure 15.1 Averaged evolution of Earth's climates at: constant Earth's precession angle $\psi = 24°$. *Curve 1* – Average for Earth surface temperature at the sea level estimated using Eq. 2.25. *Curve 2* – Effective Earth's temperature using Eq. 2.23. *Curve 3* – Earth's atmosphere greenhouse effect using Eq. 2.6. *Curve 4* – The absolute black body temperature using Eq. 2.3. using the distance of the Earth-from-the-Sun, (d) describing the increase with time of the Sun's luminosity, *S*. *Curve 5* – Average Earth's temperature on the assumption that there was no nitrogen consumption by bacteria.

bacteria and thunderstorms not existed, the current atmospheric pressure would be ≈2 atm. and Earth's average surface temperature would be over 50 °C (rather than today's 15 °C) and would exceed 70 °C at the equator. This would exceed the coagulation temperature of some vitally important proteins required for life. The conditions appropriate for life could then only be preserved on mountain summits at high latitudes. However, under conditions so extreme, the oxygen necessary for life would not be able to accumulate in sufficient amounts. In fact, should nitrogen have not been continually removed from the Earth's atmosphere, today, as in Archaean time, the Earth would be populated only by thermophilic bacteria and, maybe, primitive multicellulars.

To check the propriety of the reviewed theoretical constructions, the authors combined in Figure 15.2 (at an appropriate scale) the theoretical curve of the Earth's climate evolution and empirical curve of sea flint isotopic temperature (sea flints were deposited on the ocean floor and describe the temperature of the ocean near-bottom water (see Figure 1.3). As shown in Figure 15.2, theoretical constructions quite nicely approximate empirical data (with the accuracy of temperature determination from the sea flints and theoretical estimates). Thus, the Earth's climatic paradox is defined by a continuous decline of the atmospheric pressure. First, it occurred at the expense of bonding the carbon dioxide of the Archaean atmosphere in the carbonates of Earth sediment cover. Later, after Archaean time, it was due

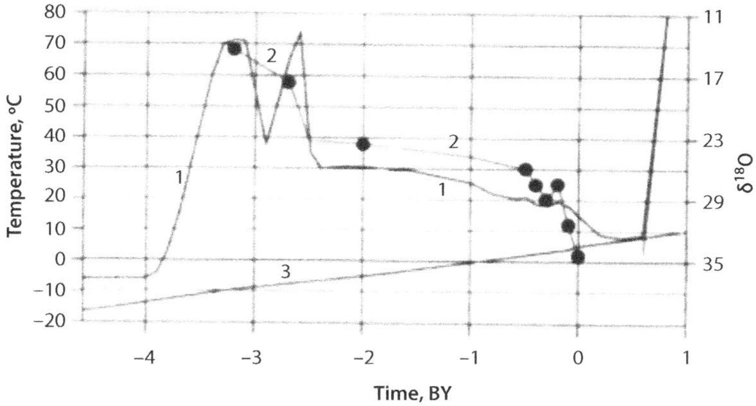

Figure 15.2 Superposition of isotopic temperature for sea flints on a theoretical curve of Earth's climate evolution: *Curve 1* – Theoretical curve of average Earth's temperature evolution on the oceanic surface (see Figure 15.1). *Curve 2* – Oxygen $\delta^{18}O$ isotope shifts in sea flints (solid black dots are estimate temperature of benthic water where the sea flints were deposited) (After Schopf, 1982). *Curve 3* – Temperature of absolutely black body at the Earth–Sun distance describing Sun luminosity (see Figure 15-3).

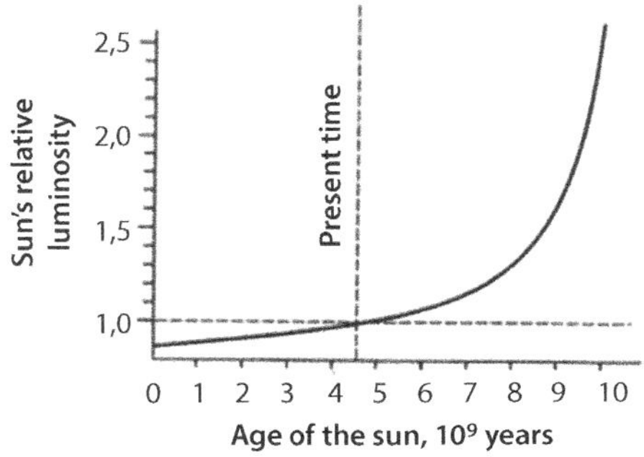

Figure 15.3 The Sun luminosity vs. time. (After Aller, 1976.)

to life-sustaining activity of the nitrogen-consuming bacteria bonding the atmospheric nitrogen in organic compounds, which then were buried with Earth's sedimentary sequences. Most of these sedimentary sequences were deposited on the global ocean floor and again pulled into the mantel through the oceanic plate subduction zones under the island arcs and continents.

Based on the oxygen $\delta^{18}O$ isotope shifts in sea flints, a substantial temperature decline in the ocean near-bottom water had already begun by the end of Proterozoic and drastically increased in Phanerozoic time. It happened exactly at the time when the first glaciations occurred on the Earth, which spread over the high-latitude areas of the World Ocean. The local benthic temperature maximum (≈200 MY ago) exactly corresponds with a warm climate epoch in the second half of Mesozoic, during the formation of the supercontinent Pangea and the total absence of traces of glaciation.

The current cooling of the Earth will not last forever because 50 to 100 MY in the future, the equilibrium between temperature decline due to (1) the bacterial removal of nitrogen from the atmosphere and (2) the temperature increases due to increasing Sun luminosity. This state of balance will likely occur ≈600 MY in the future and will be drastically disrupted by degassing of the abiogenic oxygen released at generation within the mantle of the core matter under reaction of Eq. 12.2. After that generation of the free oxygen will begin in the mantle at a rate 2.1×10^{16} g/year or 21 billion tons/ year. Should this entire supply of oxygen enter the atmosphere, its partial pressure would increase at a rate approximately 4 atm per million years. The oxygen degassing rate will be lower, but still will reach 0.02 atm per million years. This means that after 200 MY, after the beginning of oxygen degassing from the mantle, its partial pressure will have reached almost 4 atm. Earth's average surface temperature will rise to ≈76 °C due to the greenhouse effect. After another 200 MY (i.e., a billion years in the future), the oxygen pressure will have exceeded 14 atm and Earth's near-surface temperature will have risen to 110 °C.

After the beginning of endogenic oxygen degassing, life on land could not survive. Because of the low dissolved oxygen solubility in water, only in the oceans will higher life forms be able to exist. A continued warming of the surface of the Earth will result in the boiling of the ocean water (≈1.5 BY in the future) and an irreversible *greenhouse effect* will emerge with an Earth temperature ≈550 °C. Thus, even most primitive thermophilic prokaryotes will not be able to survive.

The analysis of the future Earth's climate by the authors shows that the geodynamic conditions favorable for life are not limitless. Most likely, the total duration of life on Earth is ≈4.6 BY (starting 4 BY ago and ending at ≈600 MY in the future). For the evolution of highly organized life, there is even less time. Life in the ocean has a maximum of 1.3 BY (600 MY in the past to 700 MY in the future) and on land no longer than 1 BY (400 MY in the past to 600 MY in the future).

Conclusions

"It's a paradox" (as described in Chapter 1) that despite the continual increase in ther Sun's average solar energy emissions of approximately 30% (Figure 1.10) since the beginning of the Earth's origin, the Earth's climate has cooled over the past 65 MY (see Figures 1.3 and 1.5). Furthermore, as indicated by Figure 1.7 and other temperature charts for time spans of greater than 100,000 years, the temperature of the Earth is cyclical, dependent upon the distance of the Earth from the Sun. The temperature charts prove that the average Earth's temperature is cooling rather than warming as claimed by some. Temperature charts obtained by satellite for the past 22 years show no meaningful change in temperature (Figure 1.8); this is likely due to the fact the time period is too short to display the full temperature cycle.

Chapter two develops the relationship of atmospheric pressure to temperature. In order to utilize equation for an ideal gas, the gas pressure should be less than 0.1 atmospheres. The troposphere gas pressure ranges from 1 atmosphere at sea level to about 0.3 atmosphere at its outer limits (7 to 10 miles) (see Figure 2.1). As a result, energy (heat) is primarily transferred in the troposphere by mass air convection (Figure 2.3, 2.4, and 2.5) and not radiation. Figure 2.7 demonstrates that if the Earth's atmosphere consisted of entirely carbon dioxide (at the same pressure) the Earth's surface temperature would be significantly lower and not higher as some individuals have proposed. Adding carbon dioxide to the atmosphere, lowers the amount of energy absorbed from the Sun, not increases it as proposed by many. Eq. 2.25 can be used to determine the temperature for various mixture of gases at various pressures.

As discussed in detail in the text, the development of the oceans and atmosphere have resulted in a variety of climates for the Earth over time. One must look at the long-term changes and the reason for those changes, not the short term of only a few years, as some have done, if one is to predict climate change.

Looking at the future evolution of the Earth's atmosphere, one can see a major increase in temperature that will blot out life as we know it in 600 MY. This will occur as a result of the degassing of the Earth's mantle.

Unfortunately, today the scientific community has been politicized by individuals refusing to discuss ideas different than their own. The time has come for scientific reasoning where hypotheses rise and fall by scientific facts, rather than political ones. Everyone is entitled to their own opinion, but they are not entitled to force their opinions down other people's throats; that is censorship

References and Bibliography

Abdusamatov, Kh. I., 2003, On long-term coordinated variations of luminosity activity, Sun's radius and climate; in: *Proc. of International Conference Climatic and Ecologic Aspects of the Solar Activity, St.-Petersburg*, p. 3–10.

Abdusamatov, Kh. I., 2003. On long-term coordinated activity variations of the luminosity, Sun radius and global climate. *Proc. of International Conference. Climatic and ecological aspects of the Sun activity.* SPB, p. 3–10.

Alfven, H. and Arrhenius, G., 1976. *Evolution of the Solar System.* Scientific and Technical Information Office National Aeronautics and Space Administration, Washington, DC.

Alfven, H., 1954. *On the Origin of the Solar System.* Oxford University Press, London.

Alfven, H., 1963. On the early history of the Sun and the formation of the solar system. *J. Astrophys..* 137:981–996.

Allen, D. *et al.*, 2015. Methane emissions from process equipment of natural gas production sites in the United States. *Env. Sci. & Technol.* 49:641–648.

Aller, I. H., 1971. *Atoms, Stars and Nebula.* Harvard University Press, Cambridge, MA.

Aller, L., 1976. *Atoms, Stars and Nebulae.* Moscow, Mir, 352 p.

Almukhamedov, A. I. and Medvedev, A. Va., 1982. *Sulfur Geochemistry in Basic Magma Evolution Processes*, Nauka, Moscow, 148 p.
Altshuler, I. V. and Bakanova, A. A., 1968. Metal electron structure and compressibility at high pressure. *Adv. Phys. Sci.*, 96:193–215.
Altshuler, I. V. and Sharipdzhanov, I. P., 1971b. On iron distribution in Earth and its chemical differentiation. *Izv. AN USSR, Phys. Earth*, 4:3–16.
Altshuler, I. V., 1965. Application of shock waves in physics of high pressures. *Adv. Phys. Sci.* 85:197–258.
Altshuler, I. V., Simakov, G. V., and Trunin, R. F., 1968. On the issue of the Earth's core chemical composition. *Izv. AN USSR. Phys. Earth.* **1**:3–6.
Altshuler, L. V and Sharipdzhanov, I. P., 1971a. Additive equations of state of the silicates at high pressures. *Izv. AN USSR, Phys. Earth*. 3:11–28.
Anatolyeva, A I., 1978. *Main Evolution Boundaries of the Red-Bed Formations*. Nauka, Novosibirsk, 190pp.
Anderson, Don I., 1989. *Theory of the Earth*. Blackwell Scientific Publications.
Anderson, R. N. and Hobart, V. A., 1976. The relation between heat flow, sediment thickness and age in the Eastern Pacific. *J. Geophys. Res*';, 81:2968–2989.
AP Biology, 2017. Nitrogen cycle. Khan academy, https://www.khanacademy.org/science/biology/ecology/biogeochemical-cycles/a/the-nitrogen-cycle.
Archer, D., 2010. The Global Carbon Cycle. http://globalchange.umich.edu/blobalchange1/current/lectures/kling/carbon_cycle/html
Areshev, E. G. 2003. *Oil and Gas Potential of the Marginal Seas in the Far East and SE Asia*. AVANTI, Moscow, 288pp.
Artyushkov, E. V 1968. Gravity convection in Earth's depth. *Izv. AN USSR., Phys. Earth*, 9:3–18.
Artyushkov, E. V 1970. Earth's matter density differentiation and the associated phenomena. *Izv. AN USSR, Phys. Earth*, 5:18–30.
Artyushkov, E. V and Sobolev, S. V., 1984. Physics of kimberlite magmatism; in: *Kimberlites I: Kimberlites and Related Rocks*, Elsevier, Amsterdam, pp. 309–322.
Artyushkov, E. V. and Sobolev, S. V., 1977. Mechanisms of kimberlite magma rise from depth. *Proc. AN USSR*, 236:3692–695.
Atlas of Antarctic, 1966. Vol. 1. Moscow-Leningrad. GUGK MG USSR. 236pp.
Atmosphere of Earth, 1988. *Physical Encyclopedia. Soviet Encyclopedia*, Moscow. 1:133–136.
Attenborough, D., 1984. *Life on Earth*. Moscow, Mir, 176 pp.
Avsyuk, Yu. N., 1996. *Tidal Forces and Natural Processes*. OIFZ RAN, Moscow, 188pp.
Avsyuk, Yu. N., Genshaft, Yu. S., Saltykovskii, A Yu., and Svetlosanova, Z. P. 2005. Latitudinal zonality of sedimentation areas as a manifestation of the evolution of the Earth-Moon-Sun tidal system. *Doklady Earth Sci.* 403,5:659–661.
Azbel, I. Va. and Tolstikhin, I. N., 1988. In: *Radiogenic Isotopes and Evolution of the Earth's Mantle, Crust and Atmosphere*, Geol. Inst, AN USSR, Apatity, 140 pp.
Bachall, J. N. *et al.*, 1982. Standard solar models the uncertainties in predicted capture rates of solar neutrinos. *Rev. Mod. Phys*,, 54:767–778.

Bachinsky, A. U., Putilov, V. V.and Suvorov, N. P., 1951. *Handbook of Physics*, Uchpedgiz, Moscow. 380 pp

Balanyuk, I. E. and Dongaryan, I. Sh., 1994. Role of hydrothermal methane in the formation of gas-hydrate accumulations. *Geol. Ceophys. Dev. Fields*. 3:22–28.

Balanyuk, I. E., 1981. Hydrocarbon migration from plate subduction zones to foredeeps; in: *Issues of Theoretical Geodynamics and Lithospheric Plate Tectonics*. Institute of Oceanology of AN USSR, Moscow, pp. 168–173.

Balanyuk, I. E., Matveyenkov, V V., and Sedov, A. P., 1995. Serpentinization of the oceanic upper mantle rocks as a source of the hydrocarbon formation; in: *Geologie Exploration and Utilization of the Subsurface*. RF Committee for the geology and utilization of the subsurface, Moscow, pp. 34–40.

Barazangi, M. and Dorman, J., 1968. World seismicity map of ESSA coast and geodetic survey epicenters data for 1961–1967. *Bull. Seism. Soc. Am*, 59:369–379.

Bastasch, M., 2015. Satellites: Earth is nearly in its 22nd year without global warming. http://dailycaller.com/authoir/michaelb/.

Belousov, V. V., 1954. *Fundamentals of Geotectonics*. Gosgeoltekhizdat, Moscow, 607pp.

Benioff, H. 1955b. Seismic evidence for crustal structure and tectonic activity. *Geol. Soc. Am. Spec. Paper*, 62:61–74.

Benioff, H., 1955a. Seismic evidence for crustal structure and tectonic activity. *Geol. Soc. Am. Spec. Paper*, 62:67–74.

Berzon, I. S. and Pasechnik, I. P., 1972. Dynamic parameters of the PcP wave in a case of a thin-layer model of the mantle to core transition area. *Izv. AN USSR. Phys. Earth*, 6:21–33.

Berzon, I. S., Kogan, S. D., and Pasechnik, I. P., 1968. On a possibility of constructing a thin-layer model of the shell to core transition area. *Proc. AN USSR*, 178:186–189.

Betekhtin, A. G., 1950. *Mineralogy*. Gosizdat of Geologic Literature, Moscow, 957pp.

Bijy-Duval, B., Moore, J. C., Bergen, J., *et al*., 1981. Leg 78A, Eastern Caribbean Sea. JOIDES J, VII:22l.

Birch, F., 1958. Differentiation of the mantle. *Geol. Soc. Am. Bull*., 69:4483–4486.

Blackett, P. M. S., 1961. Comparison of ancient climates with ancient latitude from rock magnetic measurements. *Proc. Roy. Soc A*, 263:1236–1248.

Boehler, R., 1993. Temperatures in the Earth's core from melting-point measurements of iron at high static pressures. *Nature*, 363:534–536.

Boehler, R., 1996. Experimental constraints on melting conditions relevant to core formation. Geochim. C03-mochim. *Acta*,_60:71109–71112.

Bogatikov, O. A., Ryabchikov, I. D., and Kononova, V. A., 1991. *Lamproites*, Nauka, Moscow.

Bogdanov, Yu. A., Lisitsin, A. P., Sagalevich, A. M., and Gurvich, E. G., 2006. *Hydrothermal Ore Genesis of the Ocean Floor*. Nauka, Moscow, 527 pp.

Bogdanov, Yu. A., Sorokhtin, O. G., Zonenshain, L. P., *et al*., 1990. *Iron-Manganese Crusts and Concretions of Underwater Mountains in the Pacific Ocean*. Nauka, Moscow, 229 pp.

Bogdanova, M. N., Yefimov, M. N., and Sorokhtin, N. O., 1992. Elements of Archaean geodynamics in the northwestern White Sea area; in: *Geodynamics and Deep Structure of the Soviet Part of the Baltic Shield.* KNC RAN, Apatity, pp. 81–9l.

Bolshakov, V. A., 2003. *New Concept of Orbital Theory of Paleo-Climate.* Taurus, Moscow, 256 pp.

Bolt, B. A., 1978. *Earthquakes.* W. H. Freeman and Company, New York.

Bolt, B., 1984. *In the depths of Earth: what the earthquakes tell us about.* Moscow, Mir, 189 pp.

Bondarenko, B. A., Garkalenko, I. A., Zhuravlev, A.V., *et al.*, 1977. New data on the structure of deep crust in the Kuril-Kamchatka trough. *Proc. AN USSR,* 234:1.

Borukayev, Ch. B., 1985. *The Structure of Precambrian and the Plate Tectonics.* Nauka, Novosibirsk, 190 pp.

Borukayev, Ch. B., *et al.*, 1977. *Precambrian of the Continents. Major Features of Tectonics.* Nauka, Novosibirsk, 264 pp.

Botte, M., 1974. *Internal Structure of Earth.* Mir, Moscow, 373 pp.

Boundless, 2017. Physical Properties of Ozone. Lumen – boundless chemistry. https://courses.lumenlearning.com/boundless-chemistry/chapter/oxygen/.

Bowen, D. Q., 1978. *Quaternary Geology: A Stratigraphic Framework for Multidisciplinary Work.* Pergamon Press, Oxford, New York, 221 pp.

Bowen, D., (1981). *Quaternary Geology, the Basis of Interdisciplinary Studies,* Mir, Moscow, 272 pp.

Brandt *et al.*, 2014. Methane leaks from North American natural gas system. *Science,* 14 Feb.

Brosge, W. P. and Tailleur, I. L., 1971. Northern Alaska petroleum provins; in: *Am Assoc Petrol. Geol. - -Mem. #15,* I. H. Cram, ed. Am. Assoc. of Petrol. Geol., pp. 68–99.

Broun, G. C. and Mussett, A. E., 1981. *The Inaccessible Earth.* Allen and Unwin, London.

Brown, J. M. and McQueen, R. G. 1986. Phase transition, Gnmeisen parameter and elasticity for shocked iron between 77 ~P A and 400 GP A. *J. Geophys. Res.,* 91, pp.

Budyko, M. I. 1974. *Climate Changes.* Gidrometeoizdat, Leningrad, 280 pp.

Budyko, M. I., 2002. Global warming; in: *Climate Changes and Their Consequences,* G. V. Menzhulin ed., Nauka, Sankt-Peterburg, pp. 7–12.

Bullard, E. C., Everett, J. E., and Smith, A. G., 1965. The fit of continents around Atlantic – A symposium on continental drift, *Phil. Trans. Roy. Soc. A,* 258:41–51.

Bullen, K. E., 1956. Seismology and the Broad Structure of the Earth's interior. Phys. Chern. *Earth,* 1:68–93.

Bullen, K. E., 1963. *An Introduction to the Theory of Seismology.* Cambridge University Press, London.

Bullen, K. E., 1973. Cores of the terrestrial planets. *Nature,* 243: 68–70.

Bullen, K. E., 1975. *The Earth's Density.* Chapman and Hall, London.

Burke, K. C. and Wilson, T., 1976. Hot spots on the Earth's surface. *Sci Am*, 235:246–257.

Bykhover, N. A., 1984. *World Mineral Resource Distribution by the Ore-Formation Epochs*. Nedra, Moscow, 576 pp.

Byshev, V. I., Neiman, V. G. and Rornanov, Yu. A., 2005. On multi-directionality of the global climatic changes on continents and in oceans. *Proc. RAN*, 400(1):98–104.

Cameron, A G. W. and Ward, W. R., 1976. The origin of the Moon. *Lunar* Sci.,7:120–132.

Cameron, A. C. W., 1973. Accumulation processes in the primitive solar nebula. *Icarus*,18:407–450.

Cameron, A. G. W. and Benz, W., 1990. The origin of the Moon and the single impact hypothesis IV *Icarus*. 92:204–216.

Cameron, A. G. W.,1972. The origin and evolution of the solar system. *Sci. Am*. 233:32–41.

Carslow, G. and Yeager, D., 1964. *Heat-Conductivity of Solid Bodies*. Moscow, Nauka.

Cestenkorn, H., 1967. On the controversy over the effect of tidal friction upon the history of the Earth-Moon system. *Icarus*. 7:2160–2167.

Chandrasekhar, S., (1961). *Hydrodynamic and Hydromagnetic Stability*. Oxford University Press, Oxford, England, Chap. VI, p.198.

Chilingar, G. V. and Khilyuk, L. F., 2007. Humans are not responsible for global warming. SPE 109202, Paper presented at SPE Technical Conference and Exhibition, Anaheim, California, 11–14 Nov., pp. 1–12.

Chilingar, G. V., 1956. Relationship between Ca/Mg ratio and geologic age. *Bull. Am. Assoc. Petrol. Geol*., 40:92256–92266.

Chilingar, G. V., Sorokhtin, O. G., and Khilyuk, L. F., 2014. Do increasing contents of methane and carbon dioxide in the atmosphere cause global warming? *Atmospheric and Climatic Sciences* (ACS), 4:819–827.

Chilingar, G. V., Sorokhtin, O. G., Khilyuk, L., 2009. Greenhouse gases and greenhouse effect. *Environ. Geology*. 58:1207–1213.

Chumakov, N. M., 1978. *Precambrian Tillites and Tilloids*. Nauka, Moscow, 2006-published by GEOS in Russia, 202 pp.

Chumakov, N. M., 2001. General direction of climatic changes on Earth over the recent 3 billion years. *Proc. RAN*. 381(5):652–655.

Chumakov, N. M., 2004. Dynamics and likely causes of climatic variations in *Late Mesozoic; in: Climate and epochs of large biosphere restructuring*, Moscow, Nauka, pg. 149–157.

Chumakov, N. M., 2004. Glacial and non-glacial climate in Pre-Cambrian. In: *Climate and epochs of large biosphere restructuring*, Moscow, Nauka, pp. 259–270.

Chumakov, N. M., 2004b. Dynamics and possible causes of climatic changes in *Late Mesozoic; in: Climate in the Epochs of Major Biospheric Changes*. Nauka, Moscow, 149–157 pp.

Chumakov, N. M., 2004c. Glaciation and non-glaciation climate in Precambrian; in: *Climate in the Epochs of Major Biospheric Changes,* Nauka, Moscow, 259–270 pp.
Chumakov, N. M., 2006. Early-Paleozoic glaciation and Biospheric events of Late Ordovician; in: *Proc. GIN, 580 – Cause-and-Effect Associations and Factors of Global Changes in Phanerozoic,* GEOC, Moscow, pp. 25–41
Chumakov, N. M., 2004a. Climatic zoning and the climate of Cretaceous period; in: *Climate in the Epochs of Major Biospheric Changes,* Nauka, Moscow, pp. 105–123.
Clemens, W., 1963. personal communication.
CMU, 2003. Oxygen cycle, Environmental Decision Making, Science and Technology, Carnegie Mellon University. http://environ.andrew.cmu.edu/m3/s4/cycleOxygen.shtml.
Coleman, R. C., 1971a. Plate tectonic emplacement of upper mantle peridotites along continental edges. *J. Geophys. Res.,* 76:51212–51222.
Coleman, R. G., 1971b. Petrological and geophysical nature of serpentinites. *Geol. Soc. Am. Bull.,* 82,4897–82,4918.
Condie, K. C., 1981. *Archean Greenstone Belts.* Elsevier, Amsterdam.
D'Aleo, JU., 2011. New rate of stratospheric photolysis questions ozone hole. https://wattsupwiththat.com/2011/01/08/new-rate-of-stratospheric-photolysis-questions-ozone-hole/
Darwin, G. H., 1898. *The Tides Kindred Phenomena in the Solar System.* John Murray, London.
Davis, T. A., Hay, W. W., Southam, J. R., and Worslev, T. R., 1977. Estimates of Cenozoic oceanic sedimentation rates. *Science,* 197:53–55.
Dawson, J. B., 1980. *Kimberlites and Their Xenoliths.* Springer-Verlag, Heidelberg.
Delingpole, J., 2015. Global Warming 'Fabrication' by NASA and NOAA. http://www.breitbart.com/london/2014/06/23/global-warming-fabricated-by-NASAand-NOAA/
Demenitskaya, R. M. and Karasik, A. M., 1971. Issue of the Arctic Ocean Origin; in: *History of the World Ocean.* Nauka, Moscow, 306 pp.
Dewey, J. and Bird, J., 1970a. Plate tectonics and geosynclines. *Tectonophysics* 10(5/6):625–638.
Dewey, J. and Bird, J., 1970b. Mountain belts and the new global tectonics. J. *Geophys. Res.* 75(14):2625–2647.
Dickinson, W. R., 1974. Subduction and oil migration. *Geology,* 2:9.
Dietz, R S., 1961. Continent and ocean basin evolution by spreading of the sea floor. *Nature,* 190:(4779):854–857.
Dietz, R S., Holden, J. C., 1972. The breakup of Pangea; in:. *Continents Adrift. – Readings from the Scientific American.* San Francisco, Freeman and Co., 1246–1254.
Dillon, R. S., 2015. *Scientific Facts and Climate Change: CO_2 and Global Cooling,* 62 pp.
Dmitriyev, L. V., 1969. On the issue of the origin of ultramafic rocks in the Indo-Oceanic ridge. *Geochemistry,* 10:326–332.

Dmitriyev, L. V., 1973. Mid-Oceanic Ridge Bedrock Geochemistry and Petrology. GEOCHI, Moscow, Abstract of PhD thesis, 45 pp.
Dmitriyev, L. V. *et al.*, 1969. Bedrock of the rift zones in the Indian Ocean Ridge and some features of their geochemistry. *Proc. AN USSR*, 185:2444–2446.
Dmitriyev, L. V. *et al.*, 2000. Hydrogen and methane formation in serpentinization of the oceanic hyperbasites and the origin of oil. *Russ. Earth Sci. Mag*, 1:11–16.
Dodd, R. T., 1986. Meteorites. *Petrology and Geochemistry*. Mir, Moscow, 384 pp.
Du Toit, A. L., 1937. Our Wandering Continents; in: *An Hypothesis of Continental Drifting*, Oliver & Boyd, London, UK.
Duffield, W. A., 1972. A naturally occurring model of global plate tectonics. *J. Geophys. Res.*, 7:2543–2555.
Dzivonsky, A. M., Hales, A. L., and Lapwood, E. R., 1975. Parametrically simple Earth models consistent with geophysical data. *Phys. Earth Planet. Inter.*, 10:12–48.
Edmond, L. M., Measures, C., McDuff, R. E. *et al.*, 1979. Ridge crest hydrothermal activity and the balances of the major and minor elements in the ocean; in the Galapagos data Earth Planet. *Sci. Lett.*, 46:1–18.
Eldrdge, C. S., Compston, W., Williams, I. S., *et al.*, 1991. Isotope evidence for the involvement of recycled sediments in diamond formation. *IBID*, 353:649–653.
Elsasser, W. M., 1963. Early History of the Earth; in: *Earth Science and Meteorites*. North-Holland, Amsterdam, pp. 1–30.
Enzler, S. M., 2015. History of the greenhouse effect and global warming. http:www.lenntech.com/greenhouse-effect/global-warming-history.htm#ixzz3fLgle620.
EPA, 2010. Methane and nitrous oxide emissions from natural sources. U. S. Environmental Protection Agency, Washington, DC, USA.
Evans, D., 2010. Is the Western Climate Establishment Corrupt, part 5: CO_2 Emissions versus Temperature? http://joannenova.com.au/2010/10/is-the-western-climate-establishment-corrupt-part-5.
Exley, R. A, Mattey, D. P., Clague, D. A, and Pillinger, C. T., 1986. Carbon isotope systematic of a mantle "hotspot": A comparison of Loihi Seamount and MORB glasses. *Earth Planet. Sci. Lett.*, 78:189–199.
Faure, G., 1977. *Principles of Isotope Geology*, 2nd ed. John Wiley and Sons, New York.
Fedonkin, M. A., 2003. The origin of the Metazoa in the light of the Proterozoic fossil record. *Paleontol. Res.*, 7:19–41.
Fedonkin, M. A., 2004. Change in the metal availability and the biosphere eukariotization in Precambrian; in: Proceedings of the Geological Institute, 565, *Present-day Issues of Geology*, Nauka, Moscow, pp. 426–447.
Fedonkin, M. A., 2008. Role of hydrogen in the evolution of metabolic systems; in: *Issues of Emergence and Evolution of the Biosphere*, House of Books "LIBROKOM", Moscow, pp. 417–437.
Fedotov *et al.*, 1969. *Earthquakes and deep structure of the south of the Kuril Island Arc*. Moscow, Nauka, 245 p.
Fenton, C. L. and Fenton, M. A., 2003. *The Rock Book*. Dover Publications, Dover, De, 432 pp.

Fenton, G. L. and Fenton, M. A., 1989. *The Fossil Book. A Record of Prehistoric Life.* Doubleday, New York.

Fischer, H., Whalen, M., Smith, J., Mastroianni, D., and Deck, B., 1999. Ice core records of atmospheric CO_2 around the last three glacial terminations. *Sci.,* 283:1712–1714.

Fish, F., 1967. Angular moment of the planets. *Icarus,* pp. 7,251–7256.

Fisher, D. E., 1987. *The Birth of the Earth.* Columbia University Press, New York.

Fisher, O., 1889. *Physics of the Earth Crust,* 2nd ed. Macmillan, London.

Fore, G., 1989. *Fundamentals of isotope geology.* Moscow, Mir, 590 p.

Forsyth, D. and Uyeda, S., 1975. On the relative importance of the driving forces of plate motion. *Geophys. J. Roy. Astr. Soc.,* 43:163–200.

Fox, S. W., 1965. A theory of macromolecular and cellular origins. *Nature,* 205:328–340.

Francheteau, J., Needham, H. D., Chourone, P., Juteau, T., *et al.,* 1979. Massive deep-sea sulfide ore deposits discovered on the East Pacific Rise. *Nature,* 277:523–528.

Friends of Science, 2017. Providing insight into climate change, https://www.friendsofscience.org/index.php?id=3

Frolova, T. I., Rudnik, G. B., and Kashintsev, G. L., 1979. Some general patterns of the oceanic magmatism and issues of the magmatic rocks origins; in: *Oceanology, Geology of the Ocean, Sediment Formation and Magmatism of the Ocean,* Nauka, Moscow, pp. 69–87.

Frolova. T. I. and Burikova, I. A., 1992. Andesite volcanism in the Earth's history. Herald MGU. *Geology,* 4:3–17.

Fryer, B. J., 1977. Rare Earth evidence in iron-formations for changing Precambrian oxidation states. *Geochim. Cosmochim. Acta.* 41:361–367.

Galimov, E. M., 1984. $^{13}C/^{12}C$ ratio in diamonds. Vertical zoning of diamond formation in the lithosphere; in: Proceedings of 27th International Geologic Congress, Section C 11. Geochemistry and Cosmochemistry, Nauka, Moscow, 11:110–123.

Galimov, E. M., 1988a. Issues of the carbon geochemistry. *Geochemistry.* 2:258–278.

Galimov, E. M., 1988b. On the ocean's emergence and evolution based on the data of $^{18}O/^{16}O$ ratio variations in Earth's sedimentary shell in geologic time, *Proc. AN USSR.* 299:4977–4981.

Galimov, E. M., 2001. Phenomenon of Life: Between Equilibrium and Linearity; in: *Origins and Principles of the Evolution.* Editorial URSS, Moscow, 256 pp.

Galimov, E. V., 1988. On the ocean emergence and evolution based on $^{18}O/^{16}O$ variation within Earth's sedimentary shell in the course of geologic time. *DAN USSR,* 299(4):977–981.

Galushkin, Yu. 1. and Ushakov, S. A. 1978. Global image of lithospheric plate instantaneous kinematics. Herald MGU. *Geology.* 4: 20–35.

Garkalenko, 1. A. and Ushakov, S. A., 1978. Earth's crust of the Kuril Region. *Soviet Geol.* 11:46–59.

Garrels, R. M. and Mackenzie, F. T., 1971. *Evolution of Sedimentary Rocks.* W.W. Norton, New York.
Gast, P. W., 1968. Upper mantle chemistry and evolution of the Earth's crust; in: *The History of the Earth's Crust*, R. A Phinney, ed., Princeton University Press, Princeton, USA, pp. 15–27.
Gast, P. W., 1972. The chemical composition of the Earth, the Moon and chondritic meteorites; in: *The Nature of the Solid Earth,* E. C. Robertson, ed., McGraw-Hill, New York, pp. 19–40.
Gavrilov, V. P., 1978. *Planet's Black Gold,* Nedra, Moscow, 190 pp.
Gavrilov, V. P., 1986. *Origin of Oil,* Nauka, Moscow, 176 pp.
Gavrilov, V. P., 2008. On possible mechanisms of natural reserves replenishment in oil and gas fields, *Geol. Oil Gas,* 1:1–8.
Geodynamic map of the Circum-Pacific region, 1985. Free-air gravity (to degree 10). Tulsa, Ok 74101, USA
Geografiska Annaler, 2010. *Physical Geography,* 92 A(3):339–351, Sept.
Gerhard, L. C., 2004. Climate change: conflict of observational science, theory and politics. *Bull. Am. Assoc. Petrol. Geol,* 88(9):1211.
Giuli, R T., 1968a. Gravitational accretion of small masses attracted from large distances as a mechanism for planetary rotation, *Icarus,* 9:186.
Giuli, R T., 1968b. On the rotation of the Earth produced by gravitational accretion of particles, *Icarus,* 8:30l.
Glebovitsky, V. A. (ed.), 2005. *Early Precambrian of the Baltic Shield,* Nauka, Sankt-Peterburg, 711 pp.
Glikson, A. Y., 1976. Stratigraphy and evolution of primary and secondary greenstone complexes. Data on the Southern hemisphere shields; in: *Early History of Earth,* B. F. Windley, ed. John Wiley and Sons, New York.
Glukhovsky, M. Z. and Moralev, V. M., 1988. Tectonic position and petrogenesis of the Aldan Shield anorthosites. *Geol.* Geophys., 4:37–43.
Glukhovsky, M. Z., 1990. *Geologic Evolution of Basements of the Ancient Platforms* (Nuclear Concept). Nauka, Moscow, 215 pp.
Glukhovsky, M. Z., Moralev, V. M., and Kuzmin, M. I., 1994. Hot belt of the early Earth and its evolution, *Geotectonics,* 5:3–15.
Goldreich, P. and Soter, S., 1966. Q in the solar system. *Icarus,* 5:2.
Goldreich, P. and Toornre, A., 1969. Some remarks on polar wandering. *J. Geophys. Res.,* 74:2555–2567.
Goldreich, P., 1966a. History of the lunar orbit. *Rev. Geophys.,* 4:411–439.
Gore, A., 2006. *An Inconvenient Truth: The Planetary Emergency of Global Warming and What We Can Do About It.* Melcher Media. 346 p.
Gorodnitsky, A. M. and Sorokhtin, O. G., 1979. Map of the lithospheric plate thickness, 0. G. Sorokhtin ed.; in: *Geophysics* of the Ocean, Vol. 2, *Geodynamics.* Nauka, Moscow, pp.181–183.
Gorodnitsky, A. M. and Sorokhtin, O. G., 1981. Map of the calculated values of heat flow through the ocean floor; in: *Issues of Theoretical Geodynamics and*

Lithospheric Plate Tectonics, 0. G. Sorokhtin, ed., USSR Institute of Oceanology, Moscow, pp. 122–128.

Gorodnitsky, A. M., 1977. Thickness of the oceanic lithosphere and maximum height of the underwater volcanoes; in: *Tectonics of the Lithospheric Plates*, A. S. Monin, ed., USSR Institute of Oceanology, Moscow, pp. 24–32.

Gorodnitsky, A. M., Zonenshain, L. P., and Mirlin, E. G., 1978. *Reconstruction of Continents' Positions in Phanerozoic.* Nauka, Moscow, 121 pp.

Govindjee and Shevela, D., 2011. Adventures with cyanobacteria: a personal perspective. Front. Plant Sci., July. https://doi.org/10.3389/fpls.2011.00028

Green, D. H. and Ringwood, A. E., 1966. Origin of the calcalkalaine igneous rock suite. *Earth Planet. Sci. Lett*, 1:307–316.

Green, D. H. and Ringwood, A. E., 1967. The genesis of basaltic magmas. *Contrib. Miner. Petrol.*, 15:103–190.

Greenberg, J., 2015. Climate scientists "fabricated" temperature data. Fox News, February 13th at 10:11 a.m.

Greenhouse effect, climate and ecosystem change, 1989. Gidrometeoizdat, Leningrad, 558 pp.

Grinenko, V. A. and Grinenko, L. N., 1974. Geochemistry of Sulphur Isotopes, Nauka, Moscow, 274 pp.

Grossman, L. and Larimer, J. W., 1974. Early chemical history of the solar system, *Reviews of Geophysics and Space Physics*, 12(1):71–97.

Gurvich, E. G. 1998. *Metal Bearing Deposits of the World Ocean*. Moscow, Nauchny, Mir, pp. 340.

Haddon, R A. W. and Bullen, K. E., 1969. An Earth model incorporating free Earth oscillation data. *Phys. Earth Planet. Inter.*, 2: l.

Haken, G., 1980. *Synergistics.* Mir, Moscow, 404 pp.

Haken, G., 1985. *Synergistics. Hierarchy of instability in self-organizing systems and devices.* Mir, Moscow, 188 pp.

Hale, C. J., 1987. Palaeomagnetic data suggest link between Archaean-Proterozoic boundary and inner-core nucleation. *Nature,* 329:6136233–6136236.

Harris, A. W. and Kaula, W. M., 1975. A co-accretional model of satellite formation, Icarus, 24:516–524.

Hartmann, W. and Davis, D., 1975. Satellite-sized planetesimals and lunar origin, *Icarus,* 24:504–515.

Hartmann, W. and Larson, S., 1967. Angular momenta of planetary bodies, *Icarus,* 7:257–260.

Hashimoto, G., Abe, Y. and Sugita, S., 2007. The chemical composition of early terrestrial atmosphere: formation of a reducing atmosphere from Cl-like material, *J. Geophy. Res:* Planets, 112(E5), http://onlinelibrary.wiley.com/doi/10.1029/2006JE002844/full

Haskell, N. A., 1935. Motion of viscous fluid under a surface load, *Physics,* 6:265–269.

Hayka, M. (1981). *Essays of Comparative Planetology*. Moscow, Nauka, 327–327 pp.

Heirtzler, J. R., Dickson, G. 0., Herron, E. M. Pitman, III, W. C. and Le Pichon, X., 1968. Marine magnetic anomalies, geomagnetic field reversals and motions of the ocean floor and continents, *J. Geophys. Res*, 73:62119–62136.

Heiskanen, W. A. and Vening Meinesz, F. A., 1958. *The Earth and Its Gravity Field.* McGraw Hill, New York, 470 pp.

Hekinian, R., Francheteau, J., Renard, V., Ballard, R. D., *et al.*, 1983. Intense hydrothermal activity at the rise axis of the East Pacific Rise near 13 N: Submersible witnesses the growth of sulfide chimney, *Mar. Geophys. Res.*, 6:11–14.

Hekinian, R., Rosendahl, B. R., Cronan, D. S., *et al.*, 1978. Hydrothermal deposits and associated basement rocks from the Galapagos spreading center, *Oceanol. Acta*, 1:3473–3482.

Hess, H. H., 1962. *History of ocean basins. Petrologic Students. A volume to honor A.F. Buddington,* Geol. Soc. Am., 599–620.

Hieb, M., 2009. Climate and the Carboniferous Period, http://www.geocraft.com/WVFossils/Carboniferous_climate.html.

Hollend, H. D., 1984. *The Chemical Evolution of the Atmosphere and Oceans.* Princeton University Press, Princeton, NJ.

Holmes, A., 1928. Radioactivity and Earth movements, *Geol. Soc. Glasgow Trans.*, 18:559–606.

Holmes, A., 1946. An estimate of the age of the Earth, *Nature,* 157:680–684.

Houtermans, F. G., 1946. Die Isotopenhaufigkeiten im naturlichen Blei und das Alter des Urans. *Natunoissenschaften.* 33(219):185–186.

Howarth, R., W., Santoro, R. L., and Ingraffes, A., 2011. Methane and the greenhouse gases footprint of natural gas from shale formations. *Climatic Change Letters.*

Hurley, P. M., 1968. Absolute abundance and distribution of Rb, K and Sr in the Earth. *Geochim. Cosmochim. Acta.,* 323:273–284.

Hutton, J., 1788. *Theory of the Earth,* Transactions Royal Society of Edinburgh, 1:209–303.

Ida, S., Canup, R. M., and Stewart, G. R., 1997. Lunar accretion an impact-generated disk. *Nature,* 389:353–357.

Ilupin, I. P. *et al.*, 1990. *Kimberlites, Reference Book.* Nedra, Moscow, 248 pp.

Imbrie, J. and Imbrie, K. P., 1979. *Ice Ages Solving the Mystery,* Hillside, New Jersey, 264 p.

Isaks, B., Oliver, L., and Sykes, L. R., 1968. Seismology and the new global tectonics. *J. Geophys. Res.* 73:185855–185899.

Ito, E. and Takahashi, E., 1989. Postspinel transformations in the system Mg_2SiO_4-Fe_2SiO_4 and some geophysical implications, *J. Geophys. Res.*, 94:10637–10646.

Jacobs, J. A., 1975. *The Earth's Core,* Academic Press, New York.

James, H., 2008. *The Guardian,* UK, June 23.

Janke, E., Emde, F., and Losch, F., (1960). *Tafeln Hoherer Funktionen.* Stuttgart, B.G. Teubner Verlage Sell scaft.

Jaques, A. L., Lewis, J. D., and Smith, C. B., 1986. The kimberlites and lamproites of Western Australia, *Geol. Sum Australia Bull.,* 32:268.
Jeffreys, H. 1959. *The Earth,* 4th ed. Cambridge, New York, 420 pp.
Jessberger, E. K., Huneke, I. C., Podosek, F. A., and Wasserburg, G. J., 1974. High resolution argon analysis of neutron-irradiated Apollo 16 rocks and separated minerals, *Proc. Fifth Lunar Sci. Conf.,* 2:1419–1449.
Joder, H. S. and Tilley, C. E., 1962. *Origin of Basalt Magmas: An Experimental Study of Natural and Synthetic Rock Systems,* Oxford University Press, Oxford.
Joly, J., 1929. *The Surface History of Earth,* Dublin.
Jordan, T. H., 1975. The present-day motions of the Caribbean Plate, *J. Geoph. Res.,* 80:32.
Kalinin, V. A. and Sergeyeva, N. A., 1977. Variations in the internal structure of Earth caused by her evolution, *Izv. AN USSR, Phys. Earth,* 5:44–51.
Kaminsky, F. V., 1984. *Diamond Bearing of NonKimberlite Igneous Rocks,* Nedra, Moscow, 173 pp.
Kapitsa, A. P. and Gavrilov, A. A., 1999. Confirmation of a hypothesis of the Antarctic ozone hole natural origin, *Proc. RAN,* 36(4):543–546.
Kaufmann, W. L III., 1979. *Planets and Moons,* W. H. Freeman and Company, San Francisco.
Kaula, W. M. and Harris, A. W., 1973. Dynamically plausible hypotheses of lunar origin, *Nature,* 245:367–369.
Kaula, W. M., 1968. *An Introduction to Planetary Physics: The Terrestrial Planets,* Wiley and Sons, New York.
Kay, M., 1971. *North American geosynclines* (Memoir 48), Geological Society of America.
Kazakov, A. V., 1939. *Phosphate formations: Origin of phosphorites and the geologic factors of deposit formation,* Proceedings of NIUF 145, Moscow, 108 pp.
Kegwin, I. D., 1996. The little ice age and medieval warm in the Sargasso Sea, *Science,* 274:1504–1508.
Keigwin, I.D., 1996. The little ice age and medieval warm period in the Sargasso sea. *Science,* 274:1504–1508
Keondzhyan, V. P. and Monin, A S., 1976. Calculation of the planetary subsurface evolution, *Izv. AN USSR Phys. Earth,* 4:3–13.
Keondzhyan, V. P. and Monin, A. S., 1975. Gravity differentiation model of the planetary subsurface, *Proc. AN USSR,* 220:4825–4828.
Khain, V. E. 2001. *Tectonics of the Continents and Oceans,* Nauchny Mir, Moscow, 606 pp.
Khain, V. E. and Bozhko, N. A., 1988. *Historical Geotectonics, Precambrian.* Nedra, Moscow, 382 pp.
Khain, V. E. and Lomidze, M. G., 1995. *Geotectonics with the Fundamentals of Geodynamics,* MGU, Publishers, Moscow, 480 pp.
Khain, V. E. and Seslavinsky, K. B., 1994. Global rhythms in the Phanerozoic endogenous activity of Earth, *Stratigr. Geol. CO/Tel.,* 2:640–663.

Khain, V. E., 1973. *On the New Global Tectonics Issues of the Global Tectonics,* Nauka, Moscow, pp. 5–26.

Khain, V. E., 1977. Particulars of the Earth's crust evolution in Early Precambrian–Real and fictitious; in: *Issues of the Early Precambrian Geology,* Nauka, Leningrad, pp. 5–12.

Khain, V. E., 2000. Issues of the Early Precambrian Tectonics, Herald MGU, Series 4. Geology, 4:13–24.

Khain, V. E., Koronovsky, N. V. and Yasarnanov, N. A, 1997. *Historical Geology,* MGU Publishers, Moscow, 448 pp.

Kharkiv, A. D., Zinchuk, N. N., and Kryuchkov. A. I., 1998. *Bedrock Diamond Deposits of the World,* Nedra, Moscow, 555 pp.

Kheskin, L. A., Frey, F. A., Schmidt, L. A, and Smith, R. H., 1968. *Rare Earths Distribution in the Lithosphere and Cosmos,* Mir, Moscow, 188 pp.

Khilyuk, L. and Chilingar, G. V., 2003. Global warming: Are we confusing cause and effect? *Energy Sources,* 25:357–370.

Khilyuk, L. F. and Chilingar, G. V., 2004. Global warming and long-term climatic changes: a progress report, *Environ Geol.,* 46:6–7

Khilyuk, L. F. and Chilingar, G. V., 2006. On global forces of nature driving the Earth's climate. Are humans involved? *Environ Geol.,* 50:899–910.

Khilyuk, L. F. and Chilingar, G., (2004). Global warming and long-term climatic changes: A progress report, *Environ. Geol.,* 46(6):7970–7979.

Khilyuk, L. F. and Chilingar, G.V., 2003. Global warming: are we confusing cause and effect? *Energy Sources,* 25:357–370.

Khramov, A. N., 1976. *Paleomagnetic characteristics of rocks; in: Physical Properties of Rocks and of Economic Deposits,* Nedra, Moscow, pp. 189–193.

Khrornov, S. P. and Petrosyants, M. A., 2001. *Meteorology and Climatology.* MGU Publishers, Moscow, 528 pp.

Kirjoittajasta, K., 2008. Harhaluuloja ilmastonmuutoka, http://www.Harhaluuoja_ilmastonmuutoksest

Kirkinsky, V. A., 1975. Decomposition of solid solutions; in: *Experiment in Mineralogy and Petrography.* Nauka, Moscow, 1735 pp.

Knauth, L. P. and Lowe, D. R., 1978. Oxygen isotopic composition of cherts from the Onverwacht Group (3.4 BY ago), South Africa, with implications for secular variations in the isotopic composition of cherts, *Geol. Soc. Am.,* 10:436.

Kolman, R. G., 1979. *Ophiolites.* Mir, Moscow, 262 pp.

Kotlyakov, V. M., 2000. *Glaciology of the Antarctic.* Moscow, Nauka, 432 pp.

Kotlyakov, V. M., 2002. *In the world of snow and ice.* Moscow, Nauka, 384 pp.

Kovalenko, V. I. *et al.,* 1987. General patterns of the magmatism evolution in Earth's history; in: *Magmatic Rocks – Magmatism Evolution in the Earth's History,* O. A. Bogatikov, ed., Nauka, 6:332–348.

Kovalev, A. A., 1972. Plate tectonics and some aspects of metallogenetic analysis, *Geol.* Ore *Deposits,* 5:90–96.

Kovalev, A. A., 1974. On the causes of a strong oil – and gas-generation in the Middle East, *Geol. Oil Gas,* 11:69–74.

Kovalev, A. A., 1978. *Mobilism End Exploration Geologic Criteria*, Nedra, Moscow, pp. 287.

Kozlov, N. E., Sorokhtin, N. O. *et al.*, 2006. *Archaean Geology of the Baltic Shield.* Nauka, Sankt-Peterburg, 329 pp.

Kozlovsky, E. A., 2002. *Russia: Raw Material Policy and the National Security*, MGU Publishers, Moscow, 856 pp.

Kratz, O. K., *et al.*, 1981. *Evolution Stages and Types in the Precambrian Crust of the Ancient Shields*, Nauka, Leningrad, 164 pp.

Kropotkin, P. N. 1961. Paleomagnetism, paleoclimates and the issue of Earth's crust major horizontal movements, *Soviet Geol.*, 5:16–38.

Kropotkin, P. N., 1958. Value of paleomagnetism for stratigraphy and geotectonics, *Bull. MOIP Geol.*, 38:457–86.

Kucheruk, E. V and Ushakov, S. A, 1985. *Plate tectonics and oil and gas occurrences* (Geophysical Analysis), Physics of Earth, VlNITI, Moscow, 200 pp.

Kuhn, T., 1977. *Structure of the Scientific Revolutions*, Progress, Moscow, 301 pp.

Kuleshov, V. N., 1986. *Isotope Composition and Origin of the Depth Carbonates*, Moscow, Nauka, 126 pp.

Kushev, V. G., 1992. Baltic and Ukraine Shields in the general structure of the East-European basement as a system of meso – and mini-plates; in: *Depth, Structure and Geodynamics of the Crystalline Shields in the European USSR*, O. A. Bogatikova, ed., Nauka, KNC AN USSR, Apatity, pp. 85–94.

Kuskov, O. L. and Pabrichnaya, O. B., 1990. Phase relations in the system FeO-MgO-SiO_2 at the transition zones-Lower mantle boundary, *Geochemistry*, 2:266–278.

Kuskov, O. L. and Panferov, A. B., 1991a. Constitution of the mantle, density, elastic properties and the mineralogy of the 400-km discontinuity, *Phys.* Earth Planet. Inter., 69:85–100.

Kuskov, O. L. and Panferov, A. B., 1991b. Phase diagrams of the FeO-MgO-SiO_2 system and the structure of the mantle discontinuities, *Phys. Chem. Miner.*, 17:642–653.

Kuznetsov, A P., 1991. Once again on the old age and formation paths of the deep-water oceanic fauna, *Izv. AN USSR Biol.*, 2:275–284.

Landau, L. D. and Lifshits, E. M., 1976. *Statistical Physics, part 1.* Nauka, Moscow, 584 pp.

Landscheidt, T., 2003. New little ice age instead of global warming? *Energy and Environment*, 14:327–350.

Larson, R. L., Pitman, W. C. III., Golovchenko, X., Cande, S. C. *et al.*, 1985. *The Bedrock Geology of the World*, Freeman, New York, 235 pp.

Lazarenko, E. K., 1971. *Course of Mineralogy.* Higher Education Publishers, Moscow, 608 pp.

Le Pichon, X., 1968. Sea-floor spreading and continental drift, *J. Geophys. Res.*, 73:123661–123697.

Le Pichon, X., Francheteau, J., and Bonnin, J., 1973. *Plate Tectonics*, Elsevier Scientific Publishing Company, Amsterdam.

Legates, D. R., 2012. Carbon dioxide and air temperature: who leads and who follows, Cornwall Alliance.

Lehmann, I., 1934. Transmission times for seismic waves for epicentral distances around 20^0, *Geodet. lnst. Skr.,* 5:44.

Lehmann, I., 1936. *"P" Publ. Bur. Cent. Seism. Int. A,* 14:87–115.

Lehmann, I., 1959. Velocities of longitudinal waves in the upper part of the Earth's mantle, *Ann. Geophys.,* 15:93–118.

Lein, A. Yu. and Sagalevich, A. M., 2000. Smokers of the Rainboe Field as area of a large-scale abiogenic methane synthesis, *Nature,* 8:44–53.

Lein, A. Yu., *et al.*, 1998. Geochemical features of gas-bearing (CH_4) deposits of a subwater mud volcano in the Norwegian Sea, *Gcochemistnj.* 3:230–249.

Levinson-Lessing, F. Yu., 1940. *Petrography,* GIZ Geological Literature, Leningrad-Moscow, 524 pp.

Lightle, K., 2008. Solar energy, albedo, and the polar regions. http:// http://beyondpenguins.ehe.osu.edu/issue/energy-and-the-polar-environment/solar-energy-albedo-and-the-polar-regions

Linde, 2017. Industrial gases: Oxygen. http://linde-gas.com/en/products_and_supply_atmospheric/oxygen.html

Lisitsin, A. P. and Vinogradov, M. E., 1982. Global patterns of life distribution in the oceans and their reflection in the composition of the ocean-floor deposits. The formation and distribution of biogenic deposits, *Izv. AN USSR. Geol.,* 4:5–24.

Lisitsin, A. P., 1974a. *Deposit Formations in the Oceans. Quantitative Distribution of the Depositional Material,* Nauka, Moscow.

Lisitsin, A. P., 1974b. Age and composition of the oceans' basalt be,. *Proc. AN USSR.* 217:3561–565.

Lisitsin, A. P., 1978. *Processes of the Oceanic Sedimentation,* Nauka, Moscow, 392 pp.

Lisitsin, A. P., 1980. Geology of the Ocean (Oceanology, Geophysics of the Ocean), Nauka, Moscow, 464 pp.

Lisitsin, A. P., 1984. Avalanche deposition, ocean level changes, hiatuses and the pelagic deposition as the global patterns; in: *Paleo-Oceanology,* E. A. Kozlovsky, ed., 3, Nauka, Moscow, pp. 3–21.

Lisitsin, A. P., *et al.*, 1990. *Hydrothermal Formations* in the *Oceanic Rift Zones,* Nauka, Moscow, 256 pp.

Lister, C. R. B., 1972. On the thermal balance of a midocean rise, *Geophys. J. Roy. Astron. Soc.,* 26:516–520.

Litvin, Yu. A., 1991. *Physicochemical Studies of the Melting of Earth's Depth Matter,* Nauka, Moscow, 312 pp.

Lobkovsky, L. I. and Kotelkin, V. D., 2000. Double-Stage Thermochemical Model of the Mantle Convection and Its Geodynamic Consequences, *Issues of Global Geodynamics,* GEOC, Moscow, pp.29–53.

Lobkovsky, L. I. and Sorokhtin, O. G., 1976. Plastic Deformations of the Oceanic Lithosphere in the Plate Subduction Zones; in: *Tectonics of the lithospheric Plates,* Oceanology Institute, AN USSR, Moscow, p. 22–52.

Lobkovsky, L. I. and Sorokhtin, O. G., 1980. Tectonics of the lithospheric plates and the origin of the tsunami earthquakes, *Proc. AN USSR,* 251:51092–51095.

Lobkovsky, L. I., 1988. *Geodynamics of the Spreading and Subduction Zones and the Double-Staged Plate Tectonics,* Nauka, Moscow, 251 pp.

Lobkovsky, L. I., 2005. Catastrophic earthquake and tsunami 12. 26.2004 in the northern Sunda island arch, Indian Ocean: Geodynamic analysis and analogy with the central Kuril Islands, *Science,* 5:253–261.

Lobkovsky, L. I., *et al.,* 2004. *Current Issues of Geotectonics and Geodynamics,* Nauchny Mir, Moscow, 612 pp.

Loper, D. E., 1975. Torque balance and energy budget for the precession ally driven dynamo, *Phys. Earth Planet. Inter.,* 11:43–60.

Luriye, A. M., 1988. *Genesis of Copper Sandstone and Slates,* Nauka, Moscow, 183pp.

Lutz, B. G., 1985. *Magmatism of Early Earth Mobile Belts.* Nauka, Moscow.

Lyubimova, E. A. 1968. *Therrnics of Earth and Moon.* Nauka, Moscow, 280 pp.

Macdonald, C. A., Abbott, A. T., and Peterson, F. 1. 1990. Volcanoes in the Sea; in: *The Geology of Hawaii.* University of Hawaii Press, Honolulu, 518 pp.

MacDonald, G. J. F., 1964. Tidal friction, *Rev. Geophys.,* 2:467–541.

Magnitsky, V.A., 1965. *Internal Structure and Physics of the Earth,* Nedra, Moscow, 379 pp.

Makkaveyev, P. M. and Sorokhtin, O. G., 2005. Experience in the reconstruction of the carbonate balance in the ancient ocean, *Oceanology,* 45:3374–3380.

Marov, M. Ya., 1986. *Planets of the Solar System.* Nauka, Moscow, 320 pp

Mason, B. and Melson, W., 1973. *Lunar Rocks,* Mir, Moscow, 165 pp.

Mason, B., 1966. *Principles of Geochemistry,* 3rd ed. Wiley and Sons, New York, 329 pp.

Mason, B., 1971. *Basics of Geochemistry,* Moscow, Nedra, 312 pp.

Mason, B., 1971. Introduction, in *Handbook of Elemental Abundances in Meteorites,* edited by B. Masson. Gordon and Breach, New York, pp 1–19.

Mason, B., 1972. The minerology of meteorites, *Meteorites,* 7:309–326.

Matthews, M. Jr., 1998. Who's afraid of CO_2? Nast. Center for Policy Analysis, http://www.ncpa.org/pub/ba256.

Matveyenkov, V. V. and Sorokhtin, O. G., 1998. Petrological features on the initial evolutionary stages of the Loihi intraplate volcanism (Hawaiian Islands, Pacific Ocean). *Oceanologv,* 38:5742–5749.

McConnell Jr., R. K., 1965. Isostatic adjustment in a layered Earth, *J. Geophys. Res.* 70:5171–5188.

McKenzie, D. and Richter, F., 1976. Convection currents in the Earth's mantle, *Sci. Am.,* 235:572–589.

Melton, C. E. and Giardini, A. A., 1974. The composition and significance of gas released from natural diamonds from Africa and Brazil, *Am. Miner,* 59(7):8775–8782.

Melton, C. E. and Giardini, A. A., 1975. Experimental results and theoretical interpretation of gaseous inclusions found in Arkansas natural diamonds, *Am. Miner.,* 60(5):6413–6417.

Menard, G U., 1966. *Geology of the Pacific Ocean Floor*, Mir, Moscow, 273 pp.
Meyen, R. V., 1971. Modern paleobotanics and the evolution theory, *Nature*, 2:48–57.
Meynen, R. V., 1981. *Traces of Indian Grasses*, Mysl, Moscow, 159 pp.
Milankovitch, M., 1930. *Mathematische Klimatehre und Astronomische Teorie der Klimaschwan Kungen, Hundbuch der Klimatologie, lA*. Gebrruder Borntraeger, Berlin, pp. 1–176.
Miller, S. L. and Urey, H. C., 1959. Organic compound synthesis on the primitive Earth. *Science*, 130:245–251.
Miller, S. L., 1959. Formation of organic compounds on the primitive Earth; in: *The Origin of Life on Earth*, F. Clark and R. C. M. Synge, eds., Pergamon, London. pp. 123–135.
Minster, J. B., Jordan, T. N., Molnar, P., and Haines, E., 1974. Numerical modeling of instantaneous plate tectonics. Geophys. *J. Roy. Astron. Soc.*, 36:541–554.
Minz, M. V., 1992. Paleogeodynamic reconstructions of Early Precambrian on the most ancient (eastern and northeastern) part of the Baltic Shield; in: *Geodynamics and Deep Structure in the Soviet Part of the Baltic Shield*, F. P. Mitrofanov, ed. KNC RAN, Apatity, pp. 34–49.
Miyashiro, A., 1976. *Metamorphism and Metamorphic Belts*, Mir, Moscow, 536 pp.
Mokhov, I. I., Bezverkhniy, V. A, Karpenko, A. A., 2003. Milankovitch cycles and evolution of the climatic regime and atmospheric composition evolution based on the ice core data from the Antarctic "Vostok" Station. *Mater Glaciol Stud.*, 95:3–8.
Mokhov, I. I., Bezverkhny, V. A. and Kapchenko, A. A., 2003. Milankovitch cycles and the evolution of climatic regime and atmosphere parameters based on the data of ice cores at Vostok station; in: *Materials of glaciological studies*, 95:3–8.
Monckton, C., 2014. Updated Global Temperature: No global warming for 17years. http://www.climatedepot.com/2014/03/04/
Monin, A. S. and Sonechkin, D. M., 2005. *Climate Fluctuations Based on Observation Data: the Triple Sun Cycle and Other Cycles.* Nauka, Moscow. 191 pp
Monin, A. S. and Sorokhtin, O. G. 1986. On the deposits suck-in to great depths under the continents. *Proc. AN USSR*, 286:3583–3586.
Monin, A. S. and Sorokhtin, O. G., 1981. On the volumetric gravity differentiation of *Earth. Proc. AN USSR*, 259:51076–51079.
Monin, A. S. and Sorokhtin, O. G., 1982a. Earth's evolution at volumetric differentiation in her depths, *Proc. AN USSR. 263*:3572–575.
Monin, A. S. and Sorokhtin, O. G., 1982b. Evolution of the oceans and Precambrian metallogeny, *Proc. AN USSR*. 264:61453–61457.
Monin, A. S. and Sorokhtin, O. G., 1982c. Heat evolution of Earth at volumetric differentiation of her depths, *Proc. AN USSR*, 266:163–167.
Monin, A. S. and Sorokhtin, O. G., 1984. Evolution of the oceans and geochemistry of the continents; in: *Paleo-oceanology*, E. A. Kozlovsky, ed., 27th IGC 3, Nauka, Moscow. pp. 22–35.
Monin, A. S., 1977. *History of Earth*, Nauka, Leningrad, 228 pp.

Monin, A. S., 1988. *Theoretical Fundamentals of Geophysical Hydrodynamics*, Gidrometeoizdat, Leningrad, 424 pp.

Monin, A. S., 1999. *Hydrodynamics of the Atmosphere, Ocean and Earth's Depths*, Gidrometeoizdat, Moscow, 524 pp.

Monin, A. S., Seidov, D. G., Sorokhtin, O. G, and Sorokhtin, Yu.O., 1987a. Numerical experiments with forms of the mantle convection. *Proc. AN USSR*. 295:51080–51083.

Monin, A. S., Seidov, D. G., Sorokhtin, O. G., and Sorokhtin, Yu. O., 1987b. Numerical modeling of the mantle convection, *Proc. AN USSR*. 295:158–163.

Monnin, E. A., Indermuhle, A., Dallenbach, J., Fluckiger, B., Stauffer, T. F., Stocker, D., Raynaud, D. and Barnoa, J. M., 2001. Atmospheric CO_2 concentrations over the last glacial termination, *Science*. 291;112–114.

Moorbath, S., 1976. Age and isotope constraints for the evolution of Archean crust; in: *The Early History of the Earth*, B. F. Windly, ed., Wiley and Sons, New York, pp. 356–366.

Moralev, V. M. and Clukhovsky, M. Z., 1985. Partial melting of metabasites and the evolution of the Precambrian lithosphere, *Proc. AN USSR*, 284:2427–2431.

Moralev, V. M., 1986. *Early Stages in the Evolution of the Continental Lithosphere*, Nauka, Moscow, 166.

Morelli, A. and Dziewonski, A., 1987. Topography of the core-Mantle boundary and lateral homogeneity of the liquid core, *Nature*, 325:6106678–6106683.

Morgan, W. J., 1968. Rises, trenches, great faults and crustal blocks, *J. Geophys. Res.*, 73:61959–61982.

Morgan, W. J., 1971. Convection plumes in the lower mantle, *Nature*. 230:42–43.

Morgan, W. J., 1972. Deep mantle convection and plate motions, *Am. Assoc. Petrol. Geol.* 56:203–213.

Munk, V., 1968. Once again: Tidal friction, *J. Roy. Astron. Soc.*, 9:4352–4375.

Murbat, S., 1980. Boundary conditions of the Archaean crust evolution based on the age and isotope data; in: *Early history of Earth*, Moscow, Mir, pp. 356–366.

Musatov, D. I. and Mezhelovsky, N. V., 1982. *Role of Riftogenic Structures in the Formation of Oil and Gas Basins and Fields*, Moscow, VIEMS, 50 pp.

Naimark, L. M. and Sorokhtin, O. G., 1977a. Density Distribution in Earth's Model with a Lherzolite Mantle Composition and Oxide-Iron Core; in: *Tectonics of the Lithospheric Plates*. USSR Institute of Oceanology, Moscow, pp. 28–41.

Naimark, L. M. and Sorokhtin, O. G., 1977b. *Energy of Earth's Gravity Differentiation. Tectonics of the Lithospheric Plates*, USSR Institute of Oceanology, Moscow, pp. 42–56.

Naimark, L. M., 1984b. Zonal melting as a gravity differentiation mechanism, *Proc. AN USSR*, 278:3580–3584.

Nairnark, L. M., 1984a. On possible gravity differentiation mechanism in a medium which separates at heating, *Proc. AN USSR*, 276:4846–4850.

Naumov, G. B., Ryzhenko, B. N. and Khodakovsky, I. L., 1971. *Reference book of thermodynamic values (for geologists)*, Moscow, Atomizdat, 240 p.

Naumov, G. V., Ryzhenko, B. N., Khodakovsky, I. L., 1971. *Handbook of Thermodynamic Values (for Geologists),* Atomizdat, Moscow, 240 pp.

Nekrasov, B. V., 1973. *Fundamentals of general chemistry*, Vol. 1. Moscow, Khimia, 656 pp.

Nesis, K. N., 2001. When life emerged in the ocean depths, *Nature*, 4:68–69.

New Global Tectonics (Plate Tectonics), 1974. Collection of Articles. Mir, Moscow, 472 pp.

NOAA, 2015. 8 reasons why climate change critics are wrong, http://energysocialnetwork.com/blog/8-reasons-why-climate-change-critics-are-wrong.

Ohtani, E. and Ringwood, A., 1984. Composition of the core. I. Solubility of oxygen in molten iron at high temperatures, *Earth Planet. Sci. Lett,* 71:185–193.

Ohtani, E., Ringwood, A., and Hibberson, W., 1984. Composition of the core II. Effect of high pressure on solubility of FeO in molten iron, *Earth Planet. Sci. Lett.,* 71:194–103.

Ohtsuki, K. and Ida, S., 1998. Planetary rotation by accretion of planetesimals with nonuniform spatial distribution formed by the planet's gravitational perturbation, *Icarus,* 131:393–420.

Oparin, A. I., 1957. Origin of Life. AN USSR Publishers, Moscow.

Opik, E. J., 1961. Tidal deformation and the origin of the Moon, *Astronom. J.*, 66:60–67.

Orlov, Yu. L., 1984. *Mineralogy of the Diamond.* Moscow, Nauka, 264 pp.

Oro, J., 1965. Investigation of organo-chemical evolution; in: *Current Aspects of Exobiology*, G. Mamikunian and M. H. Briggs, eds., Pergamon, London, pp. 13–76.

Oro, J., 1966. Prebiological organic systems; in: *The Origin of Prebiologic Systems*, S. W. Fox, ed., Academic Press New York, pp. 137–162.

Oversby, V. M. and Ringwood, A. E., 1971. Time of formation Of the Earth's core, *Nature,* 234:463–465.

Ozima, M. and Podosek, F. A., 1983. *Noble Gas Geochemistry,* Cambridge University Press, Cambridge, London, New York, Melbourne, Sydney.

Ozone, 2014. Twenty Questions and Answers About the Ozone Layer: 2014 Update. NOAA Earth System Research Laboratory. https://www.esrl.noaa.gov/.

Panelli, G, F. A., 1972. Paleontological evidence on the Earth's rotational history since early Precambrian, *Astrophys. Space Sci.* 16:212–237.

Pankov, V L. and Kalinin, V. A., 1975. Thermodynamic parameters of rocks and minerals in conditions of Earth shell. *Izv. AN USSR Phys. Earth.* 3:3–15.

Pariysky, N. N., 1960. On the effects of Earth's tides on centenarian slowdown of Earth's revolution, *Astron. J.* 37:543–555.

Parker, R. L. and Oldenburg, D. W., 1973. Thermal model of Ocean Ridges, *Nat. Phys. Sci.* 242:122137–122139.

Parkinson, W. D., 1983. *Introduction to Geomagnetism,* Scottish Academic Press, Edinburgh and London.

Perisypkin, V I., Lein, A. Yu., Bogdanov, Yu. A., and Bortnikov, N. S., 1998. Lipids in hydrothermal formations in the area of 14°45'N and 29°N of the Mid-Atlantic Ridge, *Oceanology*, p. 385.

Perry Jr., E. C. and Tan, F. C., 1972. Significance of oxygen and carbon isotope variations in early Precambrian cherts and carbonate rocks of southern Africa, *Geol. Soc. Am. Bull.*, 83:647–664.

Perry Jr., E. C., 1967. The oxygen isotope chemistry of ancient cherts, *Earth Planet. Sci. Lett.*, 3:62.

Perry Jr., E. S., 1967. The Oxygen Isotope Chemistry of Ancient Cherts, *Earth Planet Sci. Lett.* No. 3, 62 pp.

Peyve, A. V., Knipper, N.A., Shtreys, A. A. *et al.*, 1971. Ocean and the geosynclinal process, *Proc. AN USSR*, 196:3657–3659.

Peyve; A. V. 1969. Oceanic crust of the geologic past, *Geotectonics*, 4:5–23.

Piper, J. D. F., 1976. Paleomagnetic evidence for a Proterozoic supercontinent, *Phil. Trans. Roy. Soc. Lond.*, 280A:469–490.

Pitman, W. C. and Heys, J. D., 1973. Upper cretaceons spreading rates and the great transgression, *Geol. Soc. Am. Abstr.* 5:72344–72357.

Planet Venus (atmosphere, surface, internal structure), 1989. Moscow, Nauka, 482 pp.

Polyansky, O. P., 1987. Mechanism of granite-gneiss domes diapiric buoying. Hydrodynamic model. Evolutionary models of the metamorphism processes on the shields and in the folded areas, IGIG, Novosibirsk, pp. 62–64.

Ponnamperuma, C., 1965. *Abiological Synthesis of Some Nucleic Acid Constituents; in: The Origin of Prebiological Systems,* Academic Press, New York, pp. 221–236.

Pospelov, G. M., 1971. Some issues of endogenous dynamics; in: *Issues of the General and Regional Geology*, B. N. Falaleyev, ed., Nauka, Novosibirsk, pp. 20–60.

Precambrian of the Continents 1976. *North and South America*, Nauka (Siberian Branch), Novosibirsk, 240 pp.

Precambrian of the Continents 1977. Ancient Platform of Eurasia, Novosibirsk, Nauka (Siberian Branch), 312 pp.

Precambrian of the Continents, 1976a. Australia and Africa, Nauka (Siberian Branch), Novosibirsk, 224 pp.

Prigozhin, I. R. and Stengers, I., 2003. *Order from Chaos. New Dialog of Man with the Nature*, Editorial URSS, Moscow, 312 pp.

Pugin, V. A. and Khitarov, N. I., 1978. *Experimental Petrology of the Depth Magmatism*, Nauka, Moscow, 175 pp.

Pushcharovsky, Yu. M. and Pushcharovsky, D. Yu., 1999. *Geospheres of the Earth's mantle*, Geotectonics, 1:3–14.

Ranelli, G. and Fischer, B., 1984. Diffusion creep, dislocation creep, and mantle rheology, *Phys. Earth Planet. Inter.*, 34:77–84.

Real Science, 2014. IR expert speaks out after 40 years of silence; it's the water vapor stupid, not the CO2. Real Science, https://stevengoddard.wordpress.com/2014/01/25/ir-expert-speaks-out-after-40-years-of-silence-its-the-water-vapor-stupid-and-not-the-co2/

Reynolds, R. T. and Summers, A. L., 1969. Calculation on the composition of the terrestrial planets, *J. Geophys. Res.*, 74:10.

Richter, F. M., 1973. Dynamical models of see-floor spreading, *Rev. Geophys. Space Phys*, 11:223–287.

Ringwood, A. E. and Hibberson, W., 1990. The system Fe-FeO revisited, *Phys. Chern. Miner.*, 17:313–319.

Ringwood, A. E., 1962a. A model for the upper mantle, 1, *J. Geophys. Res.*, 67:857–866.

Ringwood, A. E., 1962b. A model for the upper mantle, 2, *J. Geophys. Res.*, 67:4473–4477.

Ringwood, A. E., 1966. The chemical composition and origin of the Earth; in: *Advances in Earth Sciences,* Cambridge, MIT Press, pp. 276–356.

Ringwood, A. E., 1975. *Composition and Petrology of the Earth's Mantle*, McGraw-Hill, New York.

Ringwood, A. E., 1977. Composition of the core and implications for the origin of the Earth, *Geochem. J.*, 11:111–135.

Ringwood, A. E., 1979. *Origin of the Earth and Moon*, Springer-Verlag, Berlin.

Ringwood, A. E., 1981. *Composition and petrology of the Earth's mantle*. Nedra Moscow,, 584 pp.

Ringwood, A. E., 1982. *Origins of Earth and Moon*. Nedra Moscow, 175 p.

Robinson, A. B., Baliunas, S. L., Soon, W., and Robinson, Z. W., 1998. Environmental effects of increased atmospheric carbon dioxide, *J. Am. Physicians and Surgeons*, 3:171–178.

Robinson, A. B., Robinson, N. E., and Soon, W., 2007. Environmental effects of increased atmospheric carbon dioxide, *J. Am. Physicians Surg.*, 12:79–90.

Romankevich, E. A., 1977. *Geochemistry of Organic Matter in the Ocean*, Moscow, Nauka, 256 pp.

Romankevich, E. A., 1984. *Geochemistry of Organic Matter in the Ocean*. Springer-Verlag, 334 pp.

Rona, P., 1986. *Hydrothermal Mineralization of Spreading Areas in the Oceans*, Mir, Moscow, 160 pp.

Ronov, A. B. and Yaroshevsky, A. A., 1978. Chemical composition of the Earth crust and its shells. *Earth's Tectonosphere,* Nedra, Moscow, X pp. 376–402.

Ronov. A. B., 1993. *Stratisphere or the Sedimentary Shell of Earth*, Nauka, Moscow, 144 pp.

Rosanov, A. Yu., 2003. Fossil bacteria, sedimentogenesis and early studies of the biosphere evolution, *Paleontologic Journal,* Moscow, MAIK, No.6, pp. 1–9.

Rosen, O. M., Condie, K. C., Natapov, L. M., and Nozhkin, A. D., 1994. Archaean and Early Proterozoic evolution of the Siberian Craton: in: *A Preliminary Assessment;* in: *Archaean Crustal Evolution*. Elsevier, Amsterdam, pp. 411–459.

Rossenfoss, S., 2015. Pressure to reduce methane emissions highlights the need for better monitoring, *J. Petrol. Eng.*, 67(3):46–57.

Rudkevich, M. Ya., 1974. *Paleotectonic Criteria of Oil – and Gas-Bearing*, Nedra, Moscow, 184 pp.

Runcorn, S. K., 1962. Convection currents in the Earth's mantle, *Nature,* 195:1248–1249.
Runcorn, S. K., 1963. Paleomagnetic Methods of Investigating Polar Wandering and Continental Drift, *Soc. Econ. Paleon. Min. Special Pub.*, Tulsa, 10:47–54.
Runcorn, S. K., 1965. Changes in the convection pattern in the Earth's mantle and continental drift: Evidence for a cold origin of the Earth, *Philos. Trans. Roy. Soc. Ser. A.*, 258:228–251.
Ruskol, E. L., 1960. On the origins of Moon. I. The formation of the near-Earth swarm, *Astronom. J.,* 37:4690–4702.
Ruskol, E. L., 1963. On the origins of Moon. II. Moon's growth within the near-Earth satellite swarm, *Astronorn. J.,* 40:2288–2296.
Ruskol, E. L., 1971. On the origins of Moon. III. Some issues of the near-Earth swarm dynamics, *Astron. J.,* 48:4819–4829.
Ruskol, E. L., 1975. *On the Origins of Moon,* Nauka, Moscow, 188 pp.
Safronov, V, S., 1982. Current status of the Earth's origins theory, *Izv. AN USSR Phys.Earth,* 6:5–24.
Safronov, V. S. and Vityazev, A. V., 1983. Origins of the Solar System; in: *Results of Science and Technology, Astronomy,* G. Rives, ed., VINITI, Moscow, 4:5–93.
Safronov, V. S., 1969. *Evolution of the Preplanetary Cloud and the Formation of Earth and Planets,* Nauka, Moscow, 244 pp.
Sakai, H., Marais, D., Ueda, A., and Moore, J., 1984. Concentrations and isotope ratios of carbon, nitrogen and sulfur in ocean-floor basalts, *Geochim. Cosmochim. Acta.*, 48:122433–122441.
Salop, L. I., 1973. *General Stratigraphic Scale of Precambrian,* Nedra, Leningrad, 310 pp.
Savelyeva, G. N., 1987. *Gabbro-Ultramafic Complexes of the Uralian Ophiolites and Their Analogs in the Modern Oceanic Crust,* Nauka, Moscow, 246 pp.
Savostin, L.A., Volokitina, L. P., Zonenshain, L. P., *et al.*, 1980. Paleobathymetry of the Pacific Ocean in Late cretaceous, *Oceanology,* XX:5871–5881.
Schidlowski, M., 1987. Application of stable carbon isotopes to early biochemical evolution on Earth, *Ann. Rev. Earth Planet. Sci.,* 15:47–72.
Schlichting, H., 1964. *Crenzschicht-Theorie,* Verlag, G. Braun, Karlsruhe.
Schmidt, O. Yu., 1946. New theory of the Earth's origins, *Nature,* 7:6–16.
Schmidt, O. Yu., 1948. *Four Lectures on the Theory of the Earth's Origins*, AN USSR Publishers, Moscow, 76 pp.
Schopf, T., 1980. *Paleo-oceanography,* Harvard University Press, Cambridge, MA.
Schopf, T., 1982. *Paleo-oceanology,* Moscow, Mir, 311 p.
Scientific Psychic, 2017. Atmospheric composition of the Earth, http://www.scientificpsychic.com/etc/timeline/atmosphere-composition.html
Sclater, J. G., Jaupart, C., and Galson, D., (1980). The heat flow through oceanic and continental crust and the heat loss of the Earth, *Rev. Geophys. Space Phys.,* 18:269–311.
Sclater, J., Parsons, B., and Laupart, C., 1981. Oceans and continents: Similarities and differences in the mechanism of heat transport, *J. Geophys. Res.,* 86:11535–11552.

Semikhatov, M. A., Raaben, M. E., Sergeyev, V. N., Veis, A. F., and Artemova, O. B. 1999. Biotic events and a positive isotopic anomaly of the carbonate carbon 2.3–2.06 BY ago, *Stratigr. Geol. Correl.*, 2:3–27.

Sfjps, 2017. Origin of the Earth's Atmosphere. http://www.ux1.eiu.edu/~cfjps/1400/atmos_origin.html

Shakun, J., 2012. Global warming controversial: "Temperature vs CO_2: which is the cause, and which is the effect?" http://www.abovetopsecret.com/form/thread826768/pg1.

Shemenda, A. I., 1983. Similarity criteria at mechanical modeling of tectonic processes, *Geol. Geophys.*, 10:10–19.

Shklovsky, I. S., 1975. *Stars: Their Birth, Life and Death*, Nauka, Moscow, 368 pp.

Shklovsky, I. S., 1976. *Supernovae and the Associated Issues*, Nauka, Moscow, 440 pp.

Sillitoe, R H., 1972b. Relation of metal provinces in Western America to subduction of oceanic lithosphere, *Bull. Geol. Soc. Am.*, 83:813–818.

Sillitoe, R. H., 1972a. A plate tectonic model for the origin of porphyry copper deposits, *Econ. Geol.*, 67:184–197.

Singer, S. F., 1972. Origin of the Moon by tidal capture and some geophysical consequences, *The Moon*, 5:206.

Sinitsin, 1992. Geodynamics of the Baltic shield from the position of Archean cratons and Proterozoic mobile belts; in: *Geodynamics and deep structure of the Soviet portion of the Baltic shield*, KNZ RAN, pp. 7–16.

Sinyakov, V. I., 1987. *Fundamentals of the Ore Genesis Theory*, Nedra, Leningrad, 192 pp.

Slobodkin, A. I. *et al.*, 1995. Magnetite formation by thermophilic microorganisms, *Proc. RAN*, 345(5):694–697.

Smirnov, V. I., 1982. *Geology of Commercial Deposits*, Moscow, Nedra, 669 pp.

Smirnov, V.1. 1984. Periodicity of ore formation in the geologic history; in: *Metallogeny and ore deposits*, Moscow, Nauka, pp. 3–10.

Smirnov, Ya. B., 1980. Heat Field of the USSR Territory; in: *Explanatory note to The Heat Flow and Depth Temperature Maps 1:10,000,000*, GUGK, Moscow, 150 pp.

Smith, A. G. and Briden, J. C., 1977. In: *Mesozoic and Cenozoic Paleocontinental Maps*, Cambridge University Press, London, New York, Melbourn, 63 pp.

Smith, D. E., Kolenkiewicz, R, Dunn, P. J., Robbins, J. W., *et al.*, 1990. Tectonic motion and deformation from satellite laser ranging, *J. Geophys. Res.*, 95:B1322013-B1322041.

Sobolev, R N., 2000. Types of tin deposits and their distribution in time and space. Herald of MGV series 4, *Geology*, 1:30–39.

Sobolev, V. S. and Sobolev, N. V., 1980. New evidences of the Earth crust eclogite rocks subsidence to great depths, *Proc. AN USSR*, 250:3683–3685.

Soft Schools, 2017. Oxygen facts, Soft Schools. http://www.softschools.com/facts/periodic_table/oxygen_facts/181/.

Sokolov, Yu. M. and Kratz, K. O., 1984. Metallogenic impulses of the Earth's crust endogenous activization in Precambrian; in: *Metallogeny of the USSA Early Precambrian*. Leningrad, Nauka, p. 4–14.

Sommerfeld, A., 1944. *Mechanik.* Zweite, Revidierte Auflage.
Sorokhtin O. G. (editor), 1979. *Geodynamics (oceanology, geophysics of the ocean),* Nauka, Moscow, 2:416.
Sorokhtin, N. O. 2002. Climate of Early Precambrian and the nature of the Huron glaciation, *Herald MGU,* 5:125–42.
Sorokhtin, N. O. and Sorokhtin, O. G., 1997. Continent stand level and possible nature of the Early Proterozoic glaciation, *Proc. RAN,* 354:2234–2237.
Sorokhtin, N. O., 1994. Theoretical aspects of the Early Precambrian geodynamics on the example of the continental crust formation in the NW White Sea region; in: *Geologic Mapping of Early Precambrian Complexes,* N. V Mezhelovsky, ed., Moscow, pp. 438–498
Sorokhtin, N. O., 1997. Dual-layer model of the continental crust formation in *Archaean;* in: *Abstracts of Reports at the International Conference, New Ideas in Earth Sciences,* Moscow, 1:136–137.
Sorokhtin, N. O., 2001. Evolution of the Continental Lithosphere in Early Precambrian (on the Example of the Eastern Baltic Shield), MGU, Moscow, PhD thesis, 368 pp.
Sorokhtin, N. O., 2002. Climate of Early Precambrian and the nature of the Huron glaciation, *Herald MGU,* 5:125–42.
Sorokhtin, O. G, 2006. *Evolution and Forecast of Changes in Earth's Global Climate.* Institute for Computer Studies; NITs Regular and Chaotic Dynamics, Moscow-Izhevsk, 88 pp
Sorokhtin, O. G. and Balanyuk, J. E., 1982. On the association between oil and gas occurrences and plate subduction zones, *Oceanologv,* XXII:2236–2245.
Sorokhtin, O. G. and Lobkovsky, L. J., 1976. A mechanism of pulling-in the oceanic deposits into the oceanic plate subduction zones, *lzv. AN USSR Phys. Earth,* 5:3–10.
Sorokhtin, O. G. and Sagalevich, A. M., 1994. Hydrothermal activity on the ocean-floor, *Proc. 10 RAN,* 131:151–162.
Sorokhtin, O. G. and Sorokhtin, N. O., 2005. Earth's core separation is the main planetary process controlling geologic evolution of Earth, *lzv. Sect. Earth Sci. RAEN,* 13:99–116.
Sorokhtin, O. G. and Sorokhtin, N. O., 2006. Exogenous mechanism of the diamond origin; in: *Smirnov Collection,* V J. Smirnov, ed., RAEN, Moscow, pp. 109–130
Sorokhtin, O. G. and Ushakov, S. A., 2002. *Evolution of Earth,* Moscow, MGU, 560 pp.
Sorokhtin, O. G., (1979b). Geodynamics (Oceanology, Geophysics of the Ocean, Nauka, Moscow, 2:416
Sorokhtin, O. G., (1999). Early stages in the evolution of the Earth-Moon system, *Izv. Sect. Earth Sci. RAEN,* 2:141–153.
Sorokhtin, O. G., (2005a). Bacterial nature of the Earth's glaciations, *Herald RAN,* 75:121107–121114.
Sorokhtin, O. G., 1972. Differentiation of the Earth's matter and the evolution of tectonic processes, *lzu. AN USSR. Phys. Earth,* 7:55–66.

Sorokhtin, O. G., 1973a. Correlation of the mid-oceanic ridge topography with the lithospheric plate spreading rate, *Proc. AN USSR*, 208(6):1338–1341.

Sorokhtin, O. G., 1973b. Possible formation mechanism of the regional over thrusts and geosynclinal folding, *Izu. AN USSR, Phys. Earth*, 7:55–67.

Sorokhtin, O. G., 1974. *Global Evolution of Earth,* Nauka, Moscow, 184 pp.

Sorokhtin, O. G., 1976. Tectonics of the lithospheric plates and the nature of global transgressions; in: *Issues of Paleohydrology*, M. V Astaykin, ed., Nauka, Moscow, pp. 59–69.

Sorokhtin, O. G., 1977. Contents of radioactive elements in Earth and the radiogenic energy; in: *Tectonics of the Lithospheric Plates (Sources of the Tectonic Processes Energy and the Plate Dynamics*, A. S. Monin, ed., IOAN, Moscow, pp. 7–27.

Sorokhtin, O. G., 1979a. Issues of oil generation within the plate subduction zones; in: *Oceanology, Geophysics of the Ocean*, Vol. 2, 0. G. Sorokhtin, ed., Nauka, Moscow, pp. 377–383.

Sorokhtin, O. G., 1979b. Issues of oil generation within the plate subduction zones; in: *Oceanology, Geophysics of the Ocean*, Geodynamics, 0. G. Sorokhtin, ed., Nauka, Moscow, 2:377–383.

Sorokhtin, O. G., 1981a. Possible physicochemical processes of the Earth's core formation, *Proc. AN USSR,* 198:61327–61330.

Sorokhtin, O. G., 1981b. Possible physicochemical processes of the Earth's core formation, *Proc. AN USSR,* 198:61327–61330.

Sorokhtin, O. G., 1981c. Structure of the continental lithospheric plates and the origins of kimberlites; in: *Issues of Theoretical Geodynamics and the Tectonics of the Lithospheric Plates,* A. S. Monin, ed., Oceanology Institute of AN USSR, Moscow, pp. 161–168.

Sorokhtin, O. G., 1981d. Structure of the continental lithospheric plates and the origins of kimberlites; in: *Issues of Theoretical Geodynamics and the Tectonics of the Lithospheric Plates,* A. S. Monin, ed., Oceanology Institute of AN USSR, Moscow. pp. 161–168

Sorokhtin, O. G., 1984. *Tectonics of the Lithospheric Plates-A Modern Geologic Theory*, Znaniye, Moscow, Russian Federation, 40 pp.

Sorokhtin, O. G., 1985. Tectonics of the Lithospheric Plates and the Origins of the Diamondiferous Kimberlites; in: *General and Regional Geology*, VIEMS, Moscow, 47 pp.

Sorokhtin, O. G., 1985. *Tectonics of the Lithospheric Plates and the Origins of the Diamondiferous Kimberlites, General and Regional Geology*, VIEMS, Moscow, 47 pp.

Sorokhtin, O. G., 1987. Formation of Diamondiferous Kimberlites and the Affined Rocks from a Standpoint of the Lithospheric Plate Tectonics; in: *Geodynamic Analysis and the Patterns of Formation and Distribution of Economic Deposits*, Nedra, Leningrad, pp. 92–107.

Sorokhtin, O. G., 1987. Formation of Diamondiferous Kimberlites and the Affined Rocks from a Standpoint of the Lithospheric Plate Tectonics; in: *Geodynamic Analysis and the Patterns of Formation and Distribution of Economic Deposits*, Nedra, Leningrad, pp. 92–107.

Sorokhtin, O. G., 1988. *Origins of Moon and the Initial Stages of the Earth Evolution Life of Earth,* MGU Publishers, Moscow, pp. 5–24.
Sorokhtin, O. G., 1990. Greenhouse effect of atmosphere in the geological history of Earth, *Doklady AN SSSR,* 315(3):587–592.
Sorokhtin, O. G., 2001. Greenhouse effect: a myth and reality, *Harbinger RAEN,* 1(1):6–21.
Sorokhtin, O. G., 2004. Global evolution of Earth, *Herald RAEN,* 4:43–16.
Sorokhtin, O. G., 2005a. Bacterial nature of Earth's glaciations, *Harbinger RAN,* 75(12):1107–1122.
Sorokhtin, O. G., 2005b. Earth's precession and Pleistocene climatic cycles, *Reports RAN,*
Sorokhtin, O. G., 2005c. Bacterial nature of the Earth's glaciations, *Herald* RAN. 75:121107–121114.
Sorokhtin, O. G., 2005d. Generation of the abiogenic methane and history of Earth evolution, *Oceanology,* 45:4531–454l.
Sorokhtin, O. G., 2006a. *Evolution and Forecast of Changes in Earth's Global Climate,* Institute of Computer Studies; NITs Regular and Chaotic Dynamics, Moscow-lzhevsk, 88 pp.
Sorokhtin, O. G., 2006b. Oceanic deposits-A "cradle" of diamonds, Nauka, in Russian, 4:46–56.
Sorokhtin, O. G., 2006c. Evolution and forecast of changes in the global Earth climate; in: *Regular and chaotic dynamics,* Institute of Computer Studies, Moscow.
Sorokhtin, O. G., 2007a. Life of Earth, Regular and Chaotic Dynamics? Institute of computer studies, Moscow-Izhevsk, 452 pp.
Sorokhtin, O. G., 2007b. Life of Earth; in: *Regular and Chaotic Dynamics,* Institute of computer studies, Moscow-Izhevsk, 452 pp.
Sorokhtin, O. G., 2008. Causes of changes in the Earth's global climate, *Ecol. Industry,* pp 35–40.
Sorokhtin, O. G., Chilingar, G. V. and Khilyuk, L. F. 2007. *Global Warming and Global Cooling. Evolution of Climate on Earth.* Elsevier, Amsterdam, 330 pp
Sorokhtin, O. G., Chilingar, G. V., Kbilyuk, L. F. and Gorfunkel, M. V., 2006. Evolution of the Earth's global climate. *Energy Sources,* 28:1–19
Sorokhtin, O. G., Chilingar, G. V., Khilyuk, L. F., and Gorfunkel, M., 2006. Evolution of the Earth's Global Climate, *Energy Source,* 28:1–19.
Sorokhtin, O. G., Chilingarian, G. V., and Sorokhtin, N. O., 2011. *Evolution of Earth and Its Climate (Birth, Life and Death of Earth).* Developments in Earth and Environmental Sciences 10, Elsevier, 576 pp.
Space, 2017. Earth's atmosphere. https://www.space.com/17683-earth-atmosphere.html
Tallbloke, 2015. Peak to peak solar cycle length since Dalton minimum. https://tallbloke.wordpress.com/2015/02/06/gerry-pease-sc23-24-longest-peak-to-peak-solar-cycle-length-since-dalton-minimum/ Longest.

Trenberth, K., 2017. Providing insight to climate change, Friends of Science, National Center for Atmospheric Research, U.S.A., https://www.friendsofscience.org/index.php?id=3.

UIGI, 2017. Nitrogen (N_2) Properties, Uses and Applications Nitrogen Gas and Liquid Nitrogen. http://www.uigi.com/nitrogen.html.

Venus (Atmosphere, Surface, Internal Structure), 1989. I Nauka, Moscow, 482 pp.

Vinogradov, M. E. and Lisitsin, A. P., 1981. Global regularities of life distribution in the ocean and their reflection in the composition of sea-floor deposits. Regularities of plankton and benthos distribution in the ocean, News bulletin of AN USSR, series geology, 3:5–25.

Voitkevich, G. V., Kokin, A. V., Miroshnikov, A. E. and Prokhorov, V. G., 1990. *Handbook of geochemistry.* Nedra. Moscow. 480 pp.

Vysotsky, V. I. and Golenkova, N. P., 1980. Causes of giant hydrocarbon concentrations in the Mesopotamian trough; in: *Geology and Oil and Gas Occurrences of the Foredeeps,* A. E. Kontorovich, ed., Nauka, Moscow, pp. 208–214.

Wadati, K., 1935. On the activity of deep-focus earthquakes in the Japan Island and neighborhood. *Geophys. Mag,* 8:305–326.

Walker, J. C. G., 1980. *The oxygen cycle in the natural environment and the biogeochemical cycles,* Springer-Verlag, Berlin, Feder (DEU).

Walker, J. C. G., 1974. Stability of atmospherics oxygen, *Am. J. Sci.,* 274:193–214.

Walker, J. C. G., 1977. *Evolution of the Atmosphere,* Macmillan Publishing Company, New York, 318 pp.

Wallace, E., 2009. Forced Carbonation: Stupid Easy. http://chanticleersociety.org/forums/p/843/5678.aspx

Watanabe, S., Mishima, K. and Matsuo, S., 1983. Isotopic ratios of carbonaceous materials incorporated in olivine crystals from Hualalai Volcano, Hawaii. An approach to mantle carbon, *Geochem. J.,* 17(2):95–104.

Watts, A., 2015. Despite attempts to erase it globally, "the pause" still exists in pristine US surface temperature data, https://wattsupwiththat.com/2015/06/14/despite-attempts-to-erase-it-globally-the-pause-still-exists-in-pristine-us-surface-temperature-data/

Wegener, A, 1912. Die Entstehung der Kontinente. *Geol. Rundsch,* 3:4276–4292.

Wegener, A, 1922. Die Entstehung der Kontinente und Ozeane,3 umgearb, Vieweg, Aufl. Braunschweig.

Weizer, Ya., 1980. Evolution of $^{87}Sr/^{86}Sr$ in sea water in geologic history and its importance as indicator of the Earth's crust evolution; in: *Early history of Earth,* Moscow, Mir, p. 565–575.

Welch, 2017. Twenty Questions and Answers About the Ozone Layer: 2010 Update. NOAA Earth System Research Laboratory,.https://www.esrl.noaa.gov/csd/assessments/ozone/2010/twentyquestions/

Wells, J. W., 1963. Coral growth and geochronometry, *Nature,* 197:948–950.

Wendt, H., 1956. *In Search of Adam,* Houghton Mifflin, Boston, p. 144–171.

Wilde, S. A., Vailey, J. W., Peck, W. H., and Graham, C. M., 2001. Evidence from detrital zircons for the existence of continental crust and oceans on the Earth 4.4 BY ago, *Nature,* 409:175–178.

Wilson, A T., 1960. Synthesis of macromolecules, *Nature*, 188:1007–1009.
Wilson, J. T., 1963. A possible origin of the Hawaiian Islands, *Can. J. Phys.*, 41:863–870.
Wilson, J. T., 1965. A new class of faults and their bearing on continental drift, *Nature*, 207:343–347.
Wilson, J. T., 1968. Convection tectonics: Some possible effects upon the earth's surface from the deep mantle, *Can. J. Earth Sci.*, 25:81199–81208.
Wit, M. J., 1998. On Archean granites, greenstones, cratons and tectonics: Does evidence demand a verdict? *Precambrian. Res.*, 91:181–226.
Wit, M. J., Roering, C., Hart, R. J., Armstrong, R. A., de Ronde, C. E. J., Green, R. W. E., Tredoux, M., Peberdy, E., and Hart, R. A., 1992. Formation of an Archaean Continent, *Nature*, 357:553–562.
Wood, J. A and Mitler, H. E., 1974. Origin of the Moon by a modified capture mechanism, or half than a loaf is better a whole one, *Lunar Sci.*, V:851–853.
Wyllie, P. J., 1984. Sources of granitoid magmas at convergent plate boundaries, *Phys. Earth Planet. Inter.*, 35:112–18.
Yasamanov, N. A., 2003. Modern global warming: causes and ecological consequences, *Bull. Dubna Int. University for Nature, Society, and Man.*, 1(8);12–21.
Yeremenko, N.A. and Chilingar, G. V., 1996. *Geology of Oil and Gas on the Verge of the Century*, Nauka, Moscow, 176 pp.
Yoder Jr, H. S., 1976. *Generation of Basaltic Magma*, National Academy of Sciences, Washington, DC.
Yoshii, T., 1975. Regionality of group velocities of Rayleigh waves in the Pacific and thickening of the plate, *Earth Planet. Sci. Lett.*, 25:3305–3312.
Young, R., 2017. A scientists' march on Washington is a bad idea, *New York Times*, Jan. 31. https://www.nytimes.com/2017/01/31/opinion/a-scientists-march-on-washington-is-a-bad-idea.html
Zabanbark, A. and Konyukhov, A. I., 2005. Oil and gas potential of the continental slopes of the World Ocean: tectonic aspects, *Geotectonics*, I:99–106.
Zabanbark, A., 2004. Oil and gas basin of the Bay of Bengal as a potential hydrocarbon potential for the South Asia, *Oceanology*, 44:778–783.
Zabanbark, A., 2006. Structural features and petroleum potential of the continental slopes of Gulf of Mexico (offshore USA), *Oceanology*, 46:4596–4602.
Zachos, J., *et al.*, 2001. Trends, rhythms and aberrations in global climate 65 MA to present. *Science*, 292:686–693.
Zachos, J., Pagani, M., Sloan, I., Thomas, E., and Billups, K., 2001. Trends, rhythms and aberrations in global climate 65 Ma to Present, *Science*, 292:686–693.
Zavaritsky, A. N., 1950. Introduction to the *Petrochemistry of Igneous Rocks*, 2nd ed, AN USSR Publishers, Moscow, Leningrad, 400 pp.
Zeldovich, Ya. B. and Novikov, I. D., 1971. *Theory of Gravitation and Evolution of Stars*, Nauka, Moscow, 484 pp.
Zeldovich, Ya. B. and Novikov, I. D., 1975. *Structure and Evolution of the Universe*, Nauka, Moscow, 736 pp.

Zerr, A. and Boehler, R., 1993. Melting of (Mg. Fe) SiO_4 perovskite to 625 kilobars indication of a high temperature in the lower mantle, *Science*, 262:553–555.

Zharikov, V.A., 1976. *Fundamentals of Physicochemical Petrology*, MGU Publishers, Moscow, 420 pp.

Zharkov, V. N. and Kalinin, V.A., 1968. *Equations of State of Solid Bodies at High Pressure and Temperature,* Nauka, Moscow, 312 pp.

Zharkov, V. N. and Trubitsin, V. P. 1980. *Physics of the Planetary Subsurface*, Nauka, Moscow, 448 pp.

Zharkov, V. N., 1983. *Internal Structure of Earth and Planets*, Nauka, Moscow, 415 pp.

Zharkov, V. N., et a1., 1971. *Physics of Earth and Planets, Shapes and Internal Structure,* Nauka, Moscow, 384 pp.

Zharkov, V. N., et a1., 1974. Test distribution of the Q (t) dissipative function in the Earth's shell, Izv. *AN USSR Phys. Earth,* 12:3–12.

Zonenshain, L. P. and Gorodnitsky, A. M., 1977. Paleozoic and Mesozoic reconstructions of continents and oceans, *Geotectonics,* 2:3–23.

Zonenshain, L. P, et a1., 1977. *Global Tectonics, Magmatism and Metallogeny*, Nedra, Moscow, 231 pp.

Zueva, 2015. Temperature and ozone anomalies as indicators of volcanic soot in the stratosphere. https://www.researchgate.net/publication/276834877_Temperature_and_ozone_anomalies_as_indicators_of_volcanic_soot_in_the_stratosphere.

Author Index

Subject Index

Printed and bound by CPI Group (UK) Ltd, Croydon, CR0 4YY
17/09/2026
14964241-0003